Preface

In seminars and workshops over the past fifteen years, my UWIST colleagues and I have had the good fortune to be able to discuss style with several thousand practising engineers and scientists. This book presents the essence of our discussions and some supporting statistical evidence.

I wish to acknowledge my debt to the participants in those seminars, who listened to our ideas, tried them out, and told us which were useful and which were bunk! I hope this presentation of the 'useful' category will contribute to the gradual improvement of standards of communication in science and technology.

I am well aware that pure scientists, applied scientists and engineers like to be considered distinctive groups; but it would have been tedious for readers if I had referred continually to all three groups or even just to 'engineers and scientists'. So, since the communication problems confronting professional engineers and scientists are very similar, and since this book is concerned with how to write readably, I hope readers will accept my gradual adoption of the convenient all-embracing terms 'scientists', 'scientific writing' and 'scientific reports and papers'. I have distinguished between the groups occasionally, when there seemed an important reason for doing so.

John Kirkman
Communication Studies Unit
University of Wales Institute
   of Science and Technology
Cardiff

May 1979

GOOD STYLE
for Scientific
and Engineering Writing

John Kirkman

PITMAN

PITMAN PUBLISHING LIMITED
39 Parker Street, London WC2B 5PB

Associated Companies
Pitman Publishing Pty Ltd, Melbourne
Pitman Publishing New Zealand Ltd, Wellington
Copp Clark Pitman, Toronto

Text set in IBM Journal 11/12 pt by Morgan-Westley
Printed and bound in Great Britain
at The Pitman Press, Bath

ISBN 0 273 01423 4

Contents

1 Which Style do Scientists Prefer?

It is surely axiomatic that the aim of scientific and technical writing is to transmit information accurately, quickly and economically from one person to another. Then why do so many scientists and engineers make their writing so heavily unreadable?

Obviously, their subject matter is sometimes complex and conceptually difficult; but frequently the 'unreadability' stems from use of a style which makes the reader's task much heavier than it need be.

Scientists and engineers themselves complain about the heaviness of this style. When I suggest that passive, impersonal, turgid expression is a millstone that the technical content need not carry, I am told that papers written in any other style would be unacceptable: 'It would be thrown straight back'; 'My boss wouldn't have it'; 'Editors insist that you write passively and impersonally'; 'You must make your work sound impressive'.

Always there is anxiety that *other* engineers and scientists would not accept a departure from 'traditional' style. This seems to be true whether I am talking to young men and women or to their seniors in higher management positions. The anxiety is always caused by a spectral reactionary group 'they'.

I have believed for a long time that 'they' must really be a small minority. In hundreds of round-table discussions with individuals and groups of scientists, I have found wide agreement on what constitutes the most readable, the most effective style for scientific and technical writing. But this agreement has been hedged round with anxiety about what other readers will want or expect. So this book offers not only advice but also evidence that I hope will convince anxious readers that other engineers and scientists, 'bosses', 'editors' — 'they' — will accept, even *prefer* the style advocated here.

In three separate surveys, I asked members of the Institution of Chemical Engineers, the British Ecological Society and the

Biochemical Society to give their views on six different versions of part of a scientific paper. The subject-matter in the three surveys was different, but within each survey, the subject-matter of the six versions was the same; only the style of writing was different. Members were asked to read the versions in any order and then to say which they found 'most comfortable to read, easiest to grasp and simplest to digest'. They were also invited to add comments on why they preferred some versions and rejected others.

In all three surveys, the majority of respondents voted for direct, active, judiciously personalized writing. Active style had majorities of 28 per cent, 41 per cent and 18.5 per cent over 'traditional' passive style.

The versions on which the surveys were based, the explanatory notes that accompanied them, and the responses from readers are shown in Appendices 1 to 4, pages 103-31. Readers may like to read through those appendices now, to begin their consideration of style by getting a sense of the over-all effects produced by varying tactical choices. Readers who are well aware of the difficulties caused by unskilful writing may prefer to carry straight on, into the discussion of the tactical decisions that must be made to produce readable, effective scientific writing.

2 On Style in General

2.1 Style is Choice

Style in writing is concerned with choice. Every writer has available the enormous resources of a whole language. English presents a particularly large range of choices of individual words, and of combinations of words into small and large 'structures' — idioms, phrases, clauses, sentences, paragraphs, sections, chapters. The choices we make create the 'style', which is a term covering balance, emphasis and tone.

There is no such thing as *the correct* way of expressing any idea, fact or opinion. Each writer selects the arrangement of words that he thinks will best express his intended meaning; each writer selects the arrangement that he thinks will best give the balance, emphasis and tone necessary to produce the desired response from his readers.

For example, it is possible to contend that the following sentences convey roughly the same information — the wish of the writer to warn readers not to open Valve X before it has cooled, because of the possibility of a dangerous explosion:

Valve X is not to be opened before cooling to 18°C because of the possibility of flash-back

Do not open Valve X before it has cooled to 18°C; there is a danger of flash-back

Owing to the danger of flash-back, Valve X must not be opened before it has cooled to 18°C

On no account should you open Valve X before it cools to 18°C or you may well cause flash-back

If you open Valve X before it cools to 18°C, there will be an almighty flash-back

Danger of flash-back: do not open Valve X before it cools to 18°C.

The linguistic choices made here alter the emphasis and tone of the statements, though the meaning stays the same. The factors that make one or other of the versions more effective as written communication are not matters of accuracy or clarity of technical content, nor are they matters of grammar: they are bound up with the relationship between the writer and the reader(s), and with the context in which the exchange takes place. In other words, in judging effectiveness of communication and suitability of style, we must take into account both the *accuracy* and the *propriety* of the language chosen. If this book is to offer advice on the best style for scientific and technical writing, I must begin by defining the type(s) of writing I have in mind, by defining the writing/reading contexts in which the exchanges are to take place. Only then will it be possible to discuss which tactical choices will best convey the desired meaning with the accuracy, balance, emphasis and tone required.

Broadly, I am concerned with the types of writing that are used for passing scientific and technical information between professional staff in academic, industrial and research organizations; that is, between groups whose intellectual capacities are approximately equal but whose specialist backgrounds may be very different. I am concerned, therefore, with reports, journal articles, proposals, technical memoranda, operating instructions and procedures, specifications, and support documentation for equipment. I am not suggesting tactics for explaining complex scientific and technical ideas to the general public. Mass communication poses special problems of selection and presentation which are not the concern of the average engineer or scientist. Nor am I suggesting tactics for advertising and commercial writing. Though *all* communication is in a sense an attempt to sell someone else our ideas, there is a difference in degree between the tactics needed to advocate a change from one soap powder to another and those needed to present new scientific ideas to colleagues in industry and research. I am concerned with the main writing tasks that confront professional engineers and scientists in their day-to-day work.

Success in fulfilling these tasks is not solely a matter of choosing suitable style. I have suggested already the importance of the relationship between writer and reader(s), of the total context of the exchange. In judging effectiveness of communication, it is necessary to take into account not only the language chosen but also some or all of the following:

— reader's familiarity with the subject;
— reader's attitude to the subject, to the writer, to the industrial organization or research centre concerned, to the journal carrying the text;
— reader's physical and mental state at the time of reading;
— reader's motivation;
— reader's expectations about style, organization and layout;
— writer's organization of material;
— writer's decisions about amount of detail and placing of emphasis;
— physical appearance of the text: layout, type-size and type-face used.

It is necessary to stress that the results of my surveys show *only* the influence of a number of linguistic features on the reader's responses to a text. They show the difference that can be made by choosing language that is active rather than passive, concrete rather than abstract, and 'everyday' rather than 'special': they show that the digestibility of statements is affected by both the length and the complexity of the sentences used to express them; they show the importance of using a variety of sentence types, some personal and some impersonal, instead of artificially restricting the choice to one or the other; and they stress the importance of breaking the whole statement (that is, the whole text) into units which give the reader the information he needs in convenient order and in manageable 'bites'.

The surveys do not offer evidence on *all* factors affecting readability. They simply give supporting evidence that the tactical choices recommended in this book are likely to produce an effective style suitable for specified writing tasks.

The nature of the writing task *must* be taken into account in any judgement of suitability of style. The style of a sports report, an accident report, a chemical research paper or a lyrical description of an emotional experience must be tailored to suit three elements: the subject matter, the audience, and the context. This book does not suggest which is the best style for *all* writing: it suggests which is the best style for scientific reports, papers and support documentation.

Roughly, the choices open to a writer as he searches the resources of language can be shown as a set of oppositions:

sentences:	short v. long
/	simple v. complex
vocabulary:	short v. long
	ordinary v. grandiose
	familiar v. unfamiliar
	non-technical v. technical
	concrete v. abstract
phrasing:	normal, comfortable idiomatic expression v. special stiff scientific idioms
	direct, incisive phrasing v. roundabout, verbose phrasing
verb forms:	active v. passive
	personal v. impersonal
paragraphing:	use v. non-use
punctuation:	careful use v. casual, random use

It would be a mistake to think that good writing will be produced by consistent choice from just one side of these oppositions. Examine the best writing, and you will find variety and flexibility. Variety and flexibility are key words in our discussion. My theme is going to be that too many scientific writers try to restrict their choice to formal, third-person, passive, impersonal constructions. The cumulative effect of this is a sense of monotonous, roundabout clumsiness. To produce comfortable, digestible writing, we should choose a variety of structures and vocabulary, as we do instinctively in careful conversation and description.

2.2 Sentence Length and Complexity

To be easy to digest, sentences must be reasonably short and not too complex. The reasons for this are not grammatical: they are connected with the number of items of information the reader can absorb in a single unit or 'thought'. One of the basic conventions of written English is that words written as sentences, or marked off by the major punctuation marks, the colon and semi-colon, represent complete units of information or thought: the reader is to assume that the writer wants him to take in the information — main statement and qualifications — as single units.

The number of items of information a reader can absorb within a unit depends on his familiarity with the subject-matter. But, however familiar he is with the subject, if there are too many items of information, he will be overwhelmed. It is vital, therefore, not to pack too much information into each sentence. Consider these examples:

According to the chemiosmotic hypothesis of oxidative and photosynthetic phosphorylation proposed by Mitchell (refs 1-4), the linkage between electron transport and phosphorylation occurs not because of hypothetical energy-rich chemical intermediaries as in the orthodox view, but because oxido-reduction and adenosine triphosphate (ATP) hydrolysis are each separately associated with the net translocation of a certain number of electrons in one direction and the net translocation of the same number of hydrogen atoms in the opposite direction across a relatively ion-, acid-, and base-impermeable coupling membrane (see also ref. 5).

As a result of a critical comparison of the two standard procedures for assessing the inhibiting effect of chemicals on conidial germination, namely (a) the addition of spores in aqueous suspension to a previously dried deposit of chemical, and (b) the mixing of spore suspension and chemical followed by application of aliquots to slides, it was found, using conidia of *Venturia inaequalis* (Cke) Wint., that variability between replicates was largely eliminated by the second procedure (A. N. Author, unpub.).

The rate of seepage of water through the sea bed into the suction anchor drains will depend on the nature of the soil and indicate conditions that where pores or fissures permit steady seepage it may be necessary to reduce the flow rate by pouring grout down pipes, to be discharged round the anchor skirt and sucked into the pores or fissures.

However, it was soon realized that the peculiar non-experimental nature of the economic process rendered classical methods inappropriate, largely because the variables appearing on the right-hand side of the econometrician's equations are not subject to his control but are jointly determined by the simultaneous interaction of the equations in the system, so that these variables will not be independent of the error terms in the equations and hence classical least squares methods will produce biased and inconsistent estimates of structural parameters.

Readers who are not expert in the subject-matter of these examples may think that their main difficulty stems from the unfamiliarity of the vocabulary. Of course, if we do not know what the words mean, it is difficult to make reasonable judgements of the readability of texts. But even expert readers find that sentences like these leave them desperately struggling to relate all the subsidiary information to the main statement at one reading. The extracts would have been much easier to assimilate if they had been presented as several units. Even 'inexpert' readers may be surprised to find how much more manageable the statements become, even though the subject-matter remains obscure:

The orthodox explanation of the link between electron transport and phosphorylation is that it is caused by hypothetical energy-rich chemical intermediaries. Mitchell, however, explains both oxidative and photosynthetic phosphorylation with a chemiosmotic hypothesis[1-4]. He suggests that oxido-reduction and the hydrolysis of adenosine triphosphate (ATP) are each separately associated with the net translocation of electrons and hydrogen atoms: a certain number of electrons move in one direction and the same number of hydrogen atoms move in the opposite direction across a coupling membrane that is relatively impermeable to ions, acids and bases. (See also ref. 5.)

A. N. Author compared the two standard procedures for assessing how chemicals inhibit conidial germination. The procedures are:

(a) adding spores suspended in water to a dried deposit of chemical;
(b) mixing spore suspension and chemical, and then putting aliquots on slides.

He used these on conidia of *Venturia inaequalis* (Cke.) Wint. The second procedure largely eliminated variation between replicates.

The rate at which water seeps through the sea bed into the suction anchor drains depends on the type of soil. If the

rate indicates that pores or fissures are allowing steady seepage, it may be necessary to reduce the seepage by grouting.
Grout can be poured through pipes, discharged round the
anchor skirt, and sucked into the pores or fissures.

However, it was soon realized that the peculiar non-
experimental nature of the economic process makes classical
methods inappropriate. The reason for this is that the variables appearing on the right-hand side of the econometrician's
equations are not under his control but are jointly determined
by the simultaneous interaction of the equations in the
system. Accordingly, these variables will depend on the error
terms in the equations, and classical least squares methods
will produce biased and inconsistent estimates of structural
parameters.

A common feature of scientific writing is the joining together
of strings of ideas of approximately equal weight, to form long
'compound' sentences. This is usually done with the praiseworthy aim of 'getting a good flow' in the writing. Unfortunately, it usually has the opposite effect: it produces unmanage-
able chunks which readers cannot absorb comfortably:

The degree of dependence of this pattern on bed structure
and/or on production of wetted paths through the packing
by randomly moving particles of the initial liquid is of
interest, as in any specified packing arrangement, complete
bed overload by high liquid flow-rate (pre-flooding) might be
expected to result in alteration of a flow pattern, dependent
on which paths through the packing were wetted, while
changes in bed structure might result from re-packing or
'stirring' the bed.

In comparison, much better readability is produced by the use
of comparatively simple sentences, of varying lengths and consisting of a main statement plus at most one or two qualifications:

We thought it would be interesting to know to what extent
this flow pattern depends on bed structure and to what
extent it depends on the random movement of the initial
liquid particles making wetted paths through the packing.
In any given packing arrangement, re-packing or 'stirring'
might be expected to change the structure of the bed;

complete overloading of the bed with a high liquid flow-rate (pre-flooding) might be expected to alter a flow pattern, depending on which paths through the pattern were wetted.

However, it is important not to swing to the other extreme, and to produce an irritating 'bitty' effect by producing a staccato sequence of short, simple sentences:

This flow pattern may be dependent on two factors. The first is bed structure, the second is purely random movement of the initial liquid particles producing wetted paths through the packing. It is of interest to determine to what extent flow pattern is dependent on these two factors. For a given arrangement of packing it could be expected that complete overloading of the bed by a high liquid flow-rate (pre-flooding) would alter a flow pattern. For the same arrangement, re-packing or 'stirring' the bed would be expected to cause changes in the structure of the bed.

Another tempting way to correct a cumbersome compound structure is to re-arrange the ideas into different levels of subordination, and to express them in a complex sentence structure. There are three pitfalls to avoid in doing this:

1. misleading the reader by faulty construction of the sentence;

2. disconcerting the reader by using not only subordinate clauses but also sub-subordinate clauses;

3. overwhelming the reader by having too many remarks in parenthesis.

The following examples illustrate these three faults; improved versions are given for comparison.

1. Faulty construction

Time division multiplexed systems are basically much simpler, combination and separation of channels being effected by timing circuits rather than by filters and inter-channel interference is less dependent on system non-linearities, due to the fact that only one channel is using the common communication medium at any instant in time.

(Systems multiplexed by time division are basically much simpler. The channels are combined and separated by timing circuits, not by filters, Interference between channels depends less on non-linear features of the system, because only one channel is using the common communication medium at any time.)

2. Sub-subordination

The most important conclusion to be drawn from this analysis is that, although the signal source is not continuously connected to the load, but only for a discrete interval of time during a switching period, the actual length of time depending on the value of k, the modulated function contains the original signal information $(f)(t)$.

(From this analysis, the most important conclusion is this: although the signal source is not continuously connected to the load, but only for a discrete interval of time during a switching period, the modulated function contains the original signal information $(f)(t)$. The interval during which the signal source is connected to the load depends on the value of k.)

3. Excessive parenthesis

Compared with the more traditional approach of clarification by coagulation and sand filtration the MD process seems likely to yield a product of rather better appearance and having a lower content of sulphate (because of the lower usage of coagulant) and chloride (if chlorine is not used), though with a higher phosphate content.

(Compared with the more traditional approach of clarification by coagulation and sand filtration, the MD process seems likely to yield a better-looking product. The product will have a higher phosphate content but it will contain less sulphate, because less coagulant is used; and if no chlorine is used, it will contain less chloride.)

Clearly, a key concept in producing readable writing is manageability. We must constantly keep in mind the information load that our readers will be able to accept in each sentence. Scientists often find this difficult. They are trained to be cautious in their observations and claims, so they are inclined to attach to any statement a string of conditions, qualifications or possible reservations. This frequently creates complex utterances, often with subjects well separated from main verbs. Readers just cannot comfortably hold in mind the amounts of inter-related information they are given in single statements:

> If the bicarbonate ion concentration, a_1, given by the previous equation, is differentiated with respect to the hydrogen ion concentration, H,
>
> $$\frac{da_1}{dH} = \frac{aK_1(K_1K_2 - H^2)}{(H^2 + K_1H + K_1K_2)^2}$$
>
> the numerator set equal to zero, and the resultant equation solved for H,
>
> $$H^2 = K_1K_2 = (4 \cdot 43 \times 10^{-7})(4 \cdot 69 \times 10^{-11})$$
>
> it can be shown that the maximum bicarbonate ion concentration occurs at a pH of $8 \cdot 34_1$ under ideal conditions.

> (It can be shown that, under ideal conditions, the maximum bicarbonate ion concentration occurs at a pH of $8 \cdot 34_1$. To do this, three steps are necessary:
>
> 1. to differentiate the bicarbonate ion concentration, a_1, given by the previous equation) with respect to the hydrogen ion concentration, H:
>
> $$\frac{da_1}{dH} = \frac{aK_1(K_1K_2 - H^2)}{(H^2 + K_1H + K_1K_2)^2} \;;$$
>
> 2. to set the numerator equal to zero;
> 3. to solve the resultant equation for H:
>
> $$H^2 = K_1K_2 = (4 \cdot 43 \times 10^{-7})(4 \cdot 69 \times 10^{-11}).)$$

> Reversal of the membrane potential from 15 mV, the inside of the cell being negative, to around 40 mV, the inside of the cell being positive, was achieved by replacing the X with Y.

> (The membrane potential was reversed by replacing the X with Y. From 15 mV with the inside of the cell negative, the potential became around 40 mV with the inside of the cell positive.)

It is important to control length and complexity, and at the same time to use a variety of structures to avoid monotony and gain emphasis. Compare the manageability and emphasis of the following two statements:

Completion of Supertron triggering is followed by commencement of anode run down and coupling back of the negative going waveform to the grid through V54A and a timing capacitor selected by Sw9/5, causing a 'Miller' type run down, at a speed controlled by the selected timing capacitor and RV51 setting. This run down, 'caught' at 106 volts by V52B, preventing possible jitter when the valve 'bottoms', is followed by regenerative stable state resumption by the Supertron, causing a sudden screen potential fall differentiated by C185 and R378, and triggering of the time base.

(After the Supertron has been triggered, the anode begins to run down and the negative-going waveform is coupled back to the grid through V54A and a timing capacitor (selected by Sw9/5). This gives a 'Miller' type run-down. The speed of the run-down is controlled by the timing capacitor selected and by the setting of RV51. The run-down is 'caught' at 106 volts by V52B, preventing jitter which might otherwise occur when the valve 'bottoms'. At the end of the run-down, the Supertron regeneratively resumes its stable state; this causes a sudden fall in the screen potential, which is differentiated by C185 and R378, and triggers the time-base.)

Even short sentences can be too densely packed. The shortest possible statement *may* not be the easiest for readers to digest. Consider the digestibility of the following groups of sentences:

For a given material thermal resistivity (inverse conductivity) varies linearly with electrical resistivity when either the orientation or the crystallinity vary.

(If either the orientation or the crystallinity of a material is changed, the thermal resistivity (inverse conductivity) and the electrical resistivity vary. The relationship between the resistivities is linear.)

(A change in the orientation or the crystallinity of a material changes its thermal resistivity (inverse conductivity) and its electrical resistivity. The relationship between the resistivities is linear.)

Check that with no trigger input free running does not occur at any setting of the trigger delay controls.

(Check that at all settings of the trigger delay controls, there is no free running when there is no trigger input.)

(Check that there is no free running when there is no trigger input. This must be true at all settings of the trigger delay controls.)

An approximately linearly decreasing relationship of modulus of elasticity with increasing temperature resulted.

(The result was a decreasing relationship between modulus of elasticity and increasing temperature. The relationship was approximately linear.)

(This resulted in an approximately linear decreasing relationship between modulus of elasticity and increasing temperature.)

These examples make an important point: that brevity is not to be recommended for its own sake in scientific writing. All three original sentences are short but difficult to comprehend at first reading. The alternative versions have been 'eased out' a little, and though they use a few more words, the increases in length are amply compensated for by increases in ease of comprehension.

Clarity and readability are the features of style to strive for. Usually these are achieved by economy of words and incisiveness of phrasing; but occasionally a densely packed statement needs to be loosened to make it easier for readers to absorb.

The apparent 'density' of a piece of writing is affected also by its layout. In my surveys, many readers commented on the difference made by the punctuation and paragraphing in the various versions. The effect came not only from more careful indication of the logic and relationship of ideas by careful punctuation, but also from the improved appearance of the text when there was white space in and around the blocks of type. It is not appropriate for me to take space in this book for detailed discussions of perception, typography, and make-up of publications.   Readers who would like to follow up these themes might like to look at M. A. Tinker's *Legibility of Print*[1], A. T. Turnbull and R. N. Baird's *The Graphics of Communica-*

tion: typography, layout and design[2], or M. Macdonald-Ross and E. Smith's excellent bibliography *Graphics in Text: a bibliography*[3]. Suffice it to say here that writers of scientific reports and papers should recognize that well-spaced type, well-paragraphed, with liberal use of sub-headings, greatly increases the readability of a text.

2.3 Vocabulary

It will have become obvious that there are two main features of construction that add to the inherent difficulty of the ideas we are expressing. They are:

1. the number of ideas we combine in each structural unit and the complexity with which we arrange them;
2. the weight and familiarity of the words we use to express the ideas.

The following statements are difficult to take in at first reading:

On the basis of this evidence it may be suggested that the major demand for cobalt in the nodulated legume lies in the requirement of the bacterial symbiont and that a reduction in nitrogen fixation in cobalt deficiency is in the main an indirect consequence of a reduction in bacteroid metabolism manifested by a reduced rate of proliferation...

The objective of this study was to evaluate alternative helicopter operating and support cost methodologies and approaches to assure that the degree of hardware design and logistic parameter sensitivity included in cost estimates accurately reflects actual expenditure sensitivities...

In making them easier to assimilate, it is helpful not only to break them into smaller units, but also to use more familiar terms:

This evidence suggests two conclusions: that the main demand for cobalt in the nodulated legume comes from the bacterial growth; and that a reduction in bacteroid metabolism (shown by slower reproduction) indirectly causes a lower nitrogen fixation in cobalt deficiency.

Our object was to evaluate ways of calculating the cost of

operating and supporting helicopters, so that estimates of design and support costs should match actual expenditure.

Our aim should always be to write as plainly and comfortably as possible, avoiding stiff formality of expression such as:

Ruminant alkaloid toxicities associated with several herbage plants including *Phalaris tuberosa* and *Heliotropum eropeaum L.* are largely avoidable by cobalt application to the animal (A. N. Author 1962).

In a face-to-face conversation with a visitor to his laboratory, the scientist who wrote that rather uncomfortable sentence would probably have said:

A. N. Author found that the poisonous effect of plants like *Phalaris tuberosa* and *Heliotropum eropeaum L.* can usually be stopped by giving cobalt to ruminant animals.

This statement would have been much easier to assimilate; it would have been just as accurate as the previous statement; it would therefore have been preferable in writing as well as in speech.

The preferred styles in all my surveys used, where possible, short words rather than long, ordinary words rather than grand, familiar rather than unfamiliar, non-technical rather than technical, and concrete rather than abstract.

A writer's first obligation is to accuracy of meaning; so where no other word or phrase will express his meaning accurately, he must use a long, technical, unfamiliar word — with supporting explanation, if necessary. But he should resist the temptation to use long, relatively unfamiliar technical words if there are short, familiar words available which would convey the meaning adequately. Note the important word 'adequately' in this advice. In technical literature, accuracy of meaning is paramount. Sometimes, unfamiliar technical language is essential for clear expression; in those circumstances, such language must be used; but where there is a choice, a writer should always use the short and the familiar. Why?

There are three main reasons why it is essential to choose shorter and more familiar words. One is sheer weight: shorter words have fewer syllables and will, therefore, reduce the heaviness of the writing. A second reason is the familiarity of these words. Familiar words cause the reader no hesitation.

An unfamiliar word causes a momentary pause as he works out its meaning. These momentary pauses have a cumulative effect and gradually create a sense of struggle to comprehend. The third reason is inflation. If we continually choose words that are a little bigger than they need to be, eventually the whole text will begin to seem inflated and pretentious.

It is not unduly fussy to look for opportunities to cut off avoidable syllables. Attention to the weight of sentences is particularly important in writing about science. It is often impossible to avoid heavy technical terms like *staphylococcus, epithelium, fertilization, ribonuclease,* and *austenitizing.* Since words like these *must* be used, it is especially important to use a minimum of syllables in building the rest of our sentences, so that the cumulative length and weight stay manageable. For example, write:

NOT	*BUT*
alate and apterous aphids	winged and wingless aphids
hyperbaric oxygen	high-pressure oxygen
performs a function analogous to	acts like a
represents the predominant mechanism for the loss of energy	is the main way energy is lost
occupies a juxta-nuclear position	is next to the nucleus
the samples had inadequate strength properties	the samples were not strong enough

The key principle is to stay as close as possible to plain, everyday language. Write about *oil-repellency,* not *oleophobicity*; write about *rates being doubled,* not *rates being increased by a factor of two.* Even apparently small changes, such as substituting *use* for *utilize, first* for *initial, kept* for *maintained,* are well worth making in sharpening the edge of a blunt and ponderous first draft.

2.4 Jargon

This is not a condemnation of 'jargon'. Jargon is often attacked wildly as being pompous and unnecessary language used deliberately to exclude laymen from understanding what is going on between specialists. Of course, if jargon *is* used deliberately to exclude anyone from understanding what is being said, it is indefensible.   But jargon terms — special words or special turns of phrase — are not coined in order to mystify. They are created to convey between fellow-specialists precise meanings which could be expressed otherwise only at much greater length.   In specialist contexts, they are the quickest and most efficient means of communication, and are wholly acceptable.

But the use of jargon has many attendant dangers. One is the temptation to use jargon outside its specialist context, to save the trouble of lengthier explanation. This will almost certainly *cause* trouble for readers. Here, for example, are two statements in which the writers attached special meanings to common words.   The reports in which the statements appeared were designed for readers other than specialists close to the writers:

the reaction is quantitative when the ratio of A : B is higher than 2 : 1

the block-to-block variation must be regarded as typical rather than representative

We cannot protest that these uses of *quantitative, typical* and *representative* are not permissible. In all spheres of activity, common words are used to express special meanings (for example, *hot* in radiation physics, *baffle* in engineering, *base* in chemistry). Our protest can be only that the writers used these words in special ways without recognizing that many of their readers would not 'understand the jargon'. In such situations, the writers were at fault. If there is any possibility that some readers will not understand a jargon word or phrase, the writer must explain it fully, or avoid using it altogether.

Another danger is that writers begin to write by stringing together jargon phrases, with an inevitable 'lumpy' effect:

Study of the *microclimatic elements* revealed that the *leaf growth rate* was limited more by *soil moisture content* than by *solar radiation effect* or *temperature influence* and that the use of the *time scales* based on *non-limiting factors* merely increased the *between-harvest variation* by producing *non-significant departures* from...

A third danger is that within his specialist context, a writer will be tempted to use jargon in place of perfectly adequate everyday expressions, simply to show that he is a specialist among specialists. To do this is pretentious; and it may mean that he will get so used to relying mechanically on prepacked jargon phrases that he will lose the capacity to find the simplest, most direct way of expressing his ideas. Consider the direct statements these writers could have made:

the problem appeared to be a function of the zinc content only
(= the zinc was causing the trouble (there was no mathematical relation intended))

it was observed that A exceeded B by a factor of X10
(= there was ten times as much A as B)

process can be applied to the determination of approximate amounts of
(= process can be used to find out roughly how much)

will be twofold enhanced
(= will be doubled)

Yet another danger of such unthinking use of jargon phrases is that those phrases gradually lose the precise meaning they once carried. If *of the order of* is continually used casually, when we mean only 'about', we shall rapidly deprive it of its more precise mathematical meaning:

the variation is of the order of a factor of four in each batch
(the variation is ...?... in each batch)

accuracies of the order of ±2½% can readily be achieved
(readily gives accuracies of about ±2½%)

of the order of 0·5-0·75
(?)

of the order of three months
(?)

of the order of ·2%
(?)

voltage drops of the order of magnitude of 10 V
(?)

of the order of at least tens of thousands of tons
(?)

Gradual change in the meanings attached to words is normal and inevitable in the natural evolution of a language. It is no use complaining that it is happening. We cannot legislate to stop it. My point is that scientists should resist the temptation to 'put on the style' a bit in their writing, to give it an air of scientific accuracy. To use words *not* because they are genuinely necessary but simply because they 'sound more scientific' is to hasten the decline in usefulness of those very words.

For example, in scientific writing, *parameter* is no longer recognized reliably to mean 'a line or quantity which determines a point, line, figure or quantity in a class of things': it is now used mainly to convey the general meaning 'variable':

there are three parameters to be taken into account

Peripheral no longer automatically signals 'of, relating to or forming a periphery, boundary or external surface': it is now used mainly to mean 'ancillary':

shall also require the following items of peripheral equipment

Viable no longer principally signals a biological meaning such as 'capable of maintaining separate existence, able to live in particular circumstances': it is now used mainly to mean 'possible, workable':

whether the scheme will be viable in present conditions

Inhibit no longer always means 'hinder, restrain, reduce': it now frequently means 'stop':

to inhibit this, add

When I read a statement such as 'chlorpromazine inhibits rotational behaviour' I am no longer sure whether it means that chlorpromazine *stops* or *restrains/reduces* rotational behaviour. I suggest it is wise, now, for scientists to avoid using these words altogether.

Metaphorical expressions suffer particularly from being overworked and casually used:

the field of animal nutrition was the first to shed light on a cobalt requirement in nature

H. W. Fowler's comments on the loose use of technical terms, 'popularized technicalities' he calls them, should be taken to heart.   He concludes[4]:

> any gratification they give to their users is at the cost of the harm done to the language by wearing down the points of words which, one suspects, may not always have been very sharp, even when confined to esoteric uses.

Let me reiterate my insistence on accuracy. My call for 'plain' language is not a call for distortion, gross over-simplification or loss of recognizable nuances or tones. I believe, for example, that there are occasions on which it is entirely proper to use the expression 'was fabricated from polyethylene sheeting'; to substitute the term *made* for *fabricated* might be to change the meaning. My point is that writers should resist the temptation to use terms like *fabricated* when *made* would serve very well, just because they feel that *fabricated* 'has a more scientific ring to it'. To do so is to run the risk of making the work sound not scientific but pompous.

2.5 Direct Phrasing

Inflated vocabulary has an insidious effect. Soon we begin to use two or three words where one would do:

samples have been subjected to examination by
(samples have been examined by)

combustion of this material can be accomplished in an atmosphere of oxygen
(this material can be burned in oxygen)

is in an extreme condition of corrosion
(is badly corroded)

X has a remarkable degree of stability
(X is remarkably stable)

reported that the batches experienced a colour change during storage
(reported that the batches changed colour during storage)

batch-to-batch variation which occurred with X does not appear to be exhibited to the same extent by Y
(Y does not seem to vary from batch to batch as much as X)

X is the procedure of choice from the point of view of cost
(X is the cheapest/least expensive procedure)

the fuel-can wall contours embrace a total thickness variation
of 0·001 in
(fuel-can wall thickness varies by 0·001 in)

water stress sufficient to cause cessation of leaf elongation
(sufficient to stop the leaves growing longer)

the beginning of the drought coincided with the commencement of ear emergence
(the drought began as the ears began to emerge)

one of the problems encountered with Glug is the difficulty
experienced in extruding the material out of 2 oz tubes
(it is difficult to squeeze Glug out of 2 oz tubes)

The storage facilities consist of a number of steel fabricated
cylindrical bottles. The bottles, linked by header tubes, are
sited in an area adjacent to the compressor building, the
ground having been excavated for the purpose of siting the
bottles, the earth being subsequently replaced such that the
bottles are now subterranean

(Storage consists of cylindrical steel bottles. The bottles,
linked by header tubes, are buried beside the compressor
building.)

(Storage consists of several cylindrical steel bottles. The
bottles, linked by header tubes, are sited underground near
the compressor building.)

Inexperienced writers may think that these rolling multi-
syllabic phrases help their work to sound more scientific. In
fact, most readers find roundabout style very tiresome. In my
surveys, the versions with most roundabout phrasing came
consistently bottom of the poll (pp.125-7). Compare these
versions of the same information:

Both stirring and pre-flooding affected the stable flow pattern,
but stirring had much greater effect than pre-flooding.

Our results reveal that both pre-flooding and stirring the bed
exerted a marked influence on the stable flow pattern; nonetheless, the effect of re-packing was of far greater magnitude.

It was apparent from the results that both pre-flooding and bed-stirring exerted a considerable influence on the stability of the flow pattern but that re-packing exerted a substantially higher degree of influence than pre-flooding.

As the swards were visibly different in form, the sites of the six plots in each sward were selected subjectively at first.

Selection of the six plots in each sward was based on an initial visual differentiation between the two swards.

At first, we chose the sites for the six plots in each of the two swards simply by looking at them: they were plainly different in form.

Site selection of the six plots in each sward was based on an initial visual differentiation between the sward types, whose heterogeneity in respect of morphology was macroscopically apparent.

The selection of the sites of the six plots in each sward was made on an initial visual differentiation between the two swards which owing to their different morphology were quite distinct to the naked eye.

Site selection for the six plots in each sward type was based on an initial differentiation between the two swards made visually, the swards being quite distinct to the unaided eye due to their morphological distinctiveness.

The normal, direct phrasing of the first statement of each of these sets also contrasts with another common weakness of scientific writing — the use of special stiff idioms where ordinary idioms are available. It would not seem normal if I wrote here that the readability of a text is vocabulary-dependent. That would seem an awkward turn of phrase. You would expect me to write that readability depends on the vocabulary used. But lumpy turns of phrase like *pressure-dependent* and *time-dependent* abound in scientific writing. They contribute greatly to its indigestible quality.

Please note that I am not suggesting that these 'scientific idioms' are not accurate or cannot be understood. My point is simply that they seem stiffer to the reader than the normal idioms. Other expressions that have a similar stiffening effect are adjectival constructions beginning with the participles *having* and *being,* and constructions in which adjectives are piled up in long strings in front of the nouns they are to qualify. Here are some examples of all these scientific idioms, with normal idioms for comparison:

Awkward turning of nouns

speed of diffusion is rate-determining
(speed of diffusion determines the rate)

appears to be temperature-dependent
(appears to depend on the temperature)

plain soap being used for staff handwashing
(; the staff used plain soap to wash their hands)

Participle constructions

a more efficient machine having the ability to cope with
(a more efficient machine which can cope with)

is essentially a box having an aperture covered with aluminium foil whose thickness is about 10 microns
(is essentially a box with an aperture covered with aluminium foil about 10 microns thick)

a request was received from a customer for a fluid having a tolerance to thermal oxidation in excess of that normally expected
(a customer asked for a fluid with a higher tolerance to thermal oxidation than is normally expected)

two-minute treatment of flour having an initial moisture content of 20% will give a product having all the desirable properties of flour made from steam-treated grain
(two-minute treatment of flour which has an initial moisture content of 20% will give a product with all the desirable properties of flour made from steam-treated grain)

in these unusual conditions, being the opposite to those in
(in these unusual conditions, which are the opposite to those in)

Excessive pre-qualification

is deposited as discrete characteristically shaped and textured crystals
(is deposited as discrete crystals with characteristic shape and texture)

using a pressure sensitive, low temperature curing glasscloth coating varnish
(using a glasscloth coating varnish which is sensitive to pressure and cures at low temperature)

Many participle constructions are the result of ill-organized thinking; and when that ill-organized thinking is also wrongly punctuated, readers are defeated:

the carriage and slide can be moved by means of lead screws having coarse and fine adjustments

(better: the carriage and slide can be moved by means of lead screws, and coarse and fine adjustments can be made)

(better still: coarse and fine adjustments of the carriage and slide can be made by means of lead screws)

a perspex plate is provided with the machine having concentric circles engraved on it

(better: a perspex plate is provided with the machine, having concentric circles engraved on it)

(better still: a perspex plate, with concentric circles engraved on it, is provided with the machine)

Excessive pre-qualification is frequently not only clumsy but also ambiguous:

a limited number of tests are made during warm-ups to obtain partial engine performance maps

(? partial maps of complete engine performance)

(? complete maps of part of the engine performance)

existing road system will be extended by 5200 metres of 6 m wide sprayed tar on crushed stone roads

(? roads 6 m wide, made of crushed stone sprayed with tar)

Once again, the model to follow is comfortable, serious conversation or discussion. This does not mean that it is acceptable to write in a loose, slangy, imprecise style, reflecting the slipshod habits of casual, everyday chatter. Loosely conversational English is usually not explicit enough in writing. For example, this statement might be clear enough and in an acceptable tone for casual conversation between fellow-workers in a laboratory:

> It turned out that if we exposed the plants to the hot material for longer, we knocked a bit off the measuring error.

For discussion with an overseas visitor, or for written transmission in a journal, this version would be too loose and slangy. Overseas readers might well be puzzled by the colloquialism 'it turned out that'; they might not understand the jargon term 'hot'; and they could certainly be forgiven for not understanding the slangy expression 'we knocked a bit off the error'. But let me stress that in re-phrasing this statement to make it acceptable in tone and clarity for a formal discussion or a written paper, it is not necessary to go to the other extreme from slang — to stiff phrasing and abstraction:

> It was found that measurement error decreased with increasing time of plant subjection to the radio-active material.

Perhaps the reason for use of this stiff phrasing is that writers feel they must give a 'proper' sense of formality to the account. They should recognize that excessive use of special scientific idioms and excessive striving for formality is likely to create a starchy, 'muscle-bound' atmosphere. The two versions below are direct, clear, and in more familiar phrasing; they are in no way offensively informal; they are, therefore, preferable both in serious conversation and in writing:

> When we exposed the plants to the radioactive material, the longer the exposure, the smaller the error in measurement.

> We found that the error in measurement was smaller if we exposed the plants to the radioactive material for longer.

We must pay attention constantly to the digestibility of the information we give our audience to assimilate. We would not normally dream of telling someone we had been to a store and bought a 'new green leather suede-lapelled patch-pocketed tiebelted jacket'. As a general rule, we recognize that listeners find it difficult to cope with the delivery of so many qualifiers before the main noun. So we give some qualifiers before it, and most of them after it. Usually, number adjectives and colour adjectives come before the noun and other qualifiers come after it: 'a new green leather jacket, with suede lapels, patch pockets, and a tie belt'. Next time you are tempted to pile up qualifiers before a noun, think of that (excruciating!) jacket, and spread out your adjectives accordingly:

Don't write
a mobile hopper fed compressed air operated grit blasting machine

Write
a mobile grit-blasting machine, fed from a hopper and operated by compressed air

2.6 Abstraction

One reason for excessive indulgence in roundabout phrasing is scientists' taste for abstract terms. Some writers always seem to choose abstract words rather than concrete:

— contraction of the tree stems occurred rapidly
— brings about the characteristic response of emergence
— causes stimulation of yield by the application of fertilizer to
— the thermal decomposition of the TMAH occurs rapidly at these temperatures and results in the formation of trimethylamine and methyl alcohol

In all these examples, an abstract noun and a roundabout phrase are used where active verbs could give directness and life to the statements. These direct, active versions have much more vigour and momentum:

— the tree stems contracted rapidly
— usually makes them come out
— stimulates the yield by putting fertilizer on
— the TMAH decomposes rapidly at these temperatures and forms trimethylamine and methyl alcohol

Too much indulgence in abstractions is almost as blurring as too much use of jargon. As Robert Gunning observes[5]:

It is difficult to pin down terms like 'capital disequilibrium' or 'idealistic parallelism'. Such terms are very far up the ladder of abstraction. And the farther up that ladder a word is, the less chance it has of meaning to the reader just what it did to the writer.

By a ladder of abstraction, Gunning means a series like:

> energy source
> fuel
> liquid fuel
> gasoline
> hexane

The bottom one of these is the most specific and concrete: the top is the most general and abstract. Each term has a proper place in our range of vocabulary, but that place depends on what we are trying to say. There are times when we want to generalize about a 'vehicle' or a 'dwelling' rather than to specify a particular type of car or house. In those circumstances, the abstraction is the best word to choose. But the writer who wrote 'we have lacked personnel capacity' made me stop and wonder if he meant more than 'we have been short of staff'. I had to puzzle over the statement 'the incidence of each of these items differs with different causal factors'; subsequently, I discovered that it meant simply 'each item has a different cause'. And I lost track completely because of the switch from *paint* to *system* in the statement 'the paints were examined again and it was found that the viscosity of all systems had increased'.

For effective writing, the principle is: do not go up the ladder of abstraction further than the sense requires. Do not write *fuel* when you mean specifically *hexane*; do not write *paint* at the beginning of a sentence, and change to *system* later in the same sentence. Take note of Sir Ernest Gowers' advice about the temptation to use abstractions[6]:

> Unfortunately the very vagueness of abstract words is one of the reasons for their popularity. To express one's thoughts accurately is hard work, and to be precise is sometimes dangerous. We are tempted to prefer the safer obscurity of the abstract. It is the greatest vice of present-day writing. Writers seem to find it more natural to say 'Was this the realization of an anticipated liability?' than 'Did you expect

to have to do this?': to say 'Communities where anonymity in personal relationships prevails' than 'Communities where people do not know one another'. To resist this temptation, and to resolve to make your meaning plain to your reader even at the cost of some trouble to yourself, is more important than any other single thing if you would convert a flabby style into a crisp one.

Notice the contrast between the 'crisp' style and the 'flabby' style, between the active constructions and the abstract circumlocutions, in the following pairs:

measurement of the torque is achieved by means of the plastograph
(the plastograph measures the torque)

to enable technicians to pursue printed circuit board servicing efficiently
(to enable technicians to service printed circuit boards efficiently)

Let me stress that this attack on *unnecessary* abstract jargon is not launched only by language specialists, protesting about their inability to understand what is going on in advanced studies in science and technology. My point is that much scientific and technical writing is made unnecessarily laborious for readers who specialize in the same subjects as the writers. Here is Sir George Pickering, Regius Professor of Medicine in the University of Oxford, talking about medical writing[7]:

the function of language is to convey information accurately and from one mind to another. Let us take a few examples to show how this most elementary function of language is neglected. Here is the kind of statement that nearly every editor of a medical journal would allow: 'Unilateral nephrectomy was performed'. Most zoologists and all doctors would understand this. Would a botanist or geologist or a dentist or a physicist or a mathematician? Perhaps if he had learnt Greek; but how many now have? If he had, he would be able to translate it into his native tongue as follows: 'One-sided kidney cutting was performed'. Then why this complicated language? Why not say 'One kidney was cut out'? The information conveyed is no less precise; it is more concise; and it is understandable by all. Here is another: 'The drug induced natriuresis and kaluresis'.   This is from a paper by an eminent physiologist in a first-rate physiological journal. Would a neurophysiologist understand, or the average physician, surgeon, or obstetrician? And if not, could they look up the meaning of the words? They could not. Nor, probably if they were Greek scholars, would their Greek take them all the way. I believe from the data given that this sentence means 'The drug increased the excretion, in the urine, of sodium and potassium'. This sentence, as interpreted, has the advantage that the information given is accurate and unequivocal and can be understood by all who speak English and know the elements of science. Then why did the author not say so, and why did the editors allow this sentence to pass?

I have chosen these two examples, because the authors in question are investigators of distinction, and men of culture and sensibility, and the journals that published them of the highest standing. These examples make it clear, I hope, that such writing is accepted practice, though a bad practice. Unnecessary and undesirable technical jargon is one of the bad habits of our time.

2.7 Verbs: Tense and Voice

Young scientists are not to be blamed for excessive use of abstractions, because usually they are simply trying to observe a 'rule' imposed on them by less discerning teachers than Sir George Pickering. The 'rule' is that scientific reports and papers must be written in the past tense, passive voice and third person (or impersonal). It is a bad rule. I believe it is responsible for much of the stupefying ponderousness of expression in scientific writing.

First, tense. It is rarely possible to write a scientific paper *wholly* in the past tense. Certainly, large parts of it should be. It is natural to use past tenses to state what the objectives were, what equipment was used and what procedure was followed. But from the very first sections, it is frequently necessary to talk about continuing objectives or persistent conditions, using *present* tenses; in descriptions of procedures, it is often necessary to quote general principles or 'eternal truths', using *present* tenses; and in discussions of data or results, it is usually necessary to state what the results *show* not *showed* and what you *deduce*. Here are some examples.

Natural mixture of tenses in introductions

Routine levelling on the site revealed certain discrepancies in the levels indicated on the brass-plate bench marks. There are three types of bench-marks on site: ordnance survey bench-marks stamped on walls; brass plates mounted on concrete plinths and stamped with both the height in feet above Newlyn datum and the level in feet with respect to site datum; and angle plates mounted on concrete plinths. Our brief was to review the site bench-mark system, taking the ordnance survey bench-mark at Admin Block as the site reference level, and to check all other marks with respect to this level...

The estimator engaged in concrete construction must be aware of the cost implications of a particular specification in order to draw up enquiries for ready-mixed concrete and analyse the subsequent quotations. This knowledge is part-icularly important when the use of site-mixed concrete is envisaged. The experiments described in this paper were designed to...

Natural mixture of tenses in descriptions of experimental procedures

The samples were examined by means of microscopical surface views (both sides), microscopical views of the sep-arated warps and shutes, and cross-sectional micrographs. To obtain the sectional view, the wire must be embedded in plastic, then ground and polished. This gives a profile such as is shown in Figure 1. To minimize the influence of the plastic coating, the material used was...

The horizontal diameters of both the six-segment ring of ordinary grey iron and the twelve-segment ring of SG iron were measured twice: immediately after they were built and while they were in the tail of the shield. The next measure-ments, taken five days later, showed the horizontal diameters of both rings had decreased. The deflection was of the opposite sense to the distortion of tunnel linings usually observed in clay. Normally there is an increase in horizontal diameter and a decrease in the vertical diameter, commonly known as squatting...

Natural mixture of tenses in discussions of results

The relations obtained between dry density and number of passes with the roller show that, for the moisture contents and depth of layer used, compaction increased only slightly in the cohesive soils and in the uniformly graded fine sand after four passes of the machine (Figure 3). The compaction increased considerably when the number of passes was increased from four to eight. An air content of 10% or less is considered to be satisfactory when...

Part of the H2 fraction was not retained on QAE-Sephadex at pH 8·3. Analysis of this fraction by isoelectric focusing and SDSP gel electrophoresis showed that it was heterogeneous, but contained predominantly low-molecular-weight species of protein, ..., ..., and . . . .. These species are rich in glycine, serine, and glutamic acid, and contain relatively low amounts of basic amino acids. Therefore, it was assumed that the basic nature of these unretained proteins resulted from extensive amidation of the free carboxyl groups...

In the conclusions section of a paper, it is natural to make *present-tense* statements about the position that has been reached. In the recommendations, it is essential to state what should be done in the *future*. It just is not possible to write wholly in the past tense.

It *is* possible to write papers almost entirely in the passive voice; but it is usually distorting and tortuous to do so. You run the risk of unintentionally changing your emphasis and meaning, and of increasing the verbosity of your text.

'Voice' is a grammatical term used to describe the possibility we have in English of viewing the action of a sentence in two ways without changing the facts reported:

John designed the laboratory (active)
The laboratory was designed by John (passive)

Acid-treatment removed the rust (active)
The rust was removed by acid-treatment (passive)

The subject position in a sentence is where we normally place (and look for) the theme or focus of that sentence. The active sentences above make statements about *John* and *acid-treatment*: the passive sentences make statements about the *laboratory* and *rust*. By using the passive construction, we move the centre of interest from the 'performer' to the 'undergoer' of the action. Though the information remains essentially the same, the emphasis is different.

The passive voice is a valuable feature of the English language and one widely used in all forms of writing and speaking, not just in science. But it is vital to note that the active and passive voices are used for specific purposes, to create deliberate balance or emphasis in a statement. They should not be interchanged arbitrarily or haphazardly.

To make an arbitrary change from 'Two thin struts linked the plates to the rig' to 'The plates were linked to the rig by two thin struts' is to change the emphasis from the *struts* to the *plates*. The following four statements are not exact equivalents. The choice of an active version or a passive version should not be dictated by an arbitrary rule of style:

— Schultz plate adjustment produced absolutely parallel loading
— Absolutely parallel loading was achieved by adjusting the Schultz plates
— Adjustment of the Schultz plates produced absolutely parallel loading
— The Schultz plates were adjusted to achieve absolutely parallel loading

A skilful writer increases the precision with which he conveys his meaning by deliberately moving from active to passive. An unskilled writer, moving from active to passive simply to satisfy a dimly understood 'rule', frequently decreases the precision with which he puts his thoughts into words.

Shift from active to passive can do more than just change emphasis or balance: sometimes it changes meaning. This happens when the verb that is converted from active to passive contains one of the auxiliary verbs such as *can, will, would,* or *might.* In an active sentence, *can* is usually taken to mean 'is able to'; in a passive sentence, it is usually taken to mean 'is possible'. Consequently, to change 'The wires can support the sheet' to 'The sheet can be supported by the wires' is to signal a different meaning to the reader. 'A three-port matrix can

represent the whole transformation' should not be lightly interchanged with 'The whole transformation can be represented as
a three port matrix'. 'Thermostatic valves will regulate the
room circuit' does not necessarily signal the same information
as 'The room circuit will be regulated by thermostatic valves'.
Active and passive constructions should not be arbitrarily interchanged: blanket imposition of passive forms may force writers
into inaccuracy.

Not only does the passive enable us to move the performer
or agent to a subordinate position and give more stress to a
more important object or activity; it also enables us to omit the
agent altogether when the agent is not relevant or not known.

It is this ability that makes the passive construction attractive
in scientific writing. Scientific writers normally wish to emphasize the procedures and findings of their work, not the people
who did it. They want accounts of scientific work to be as
detached or impersonal as possible. Use of passive constructions
achieves this.

In descriptions of experimental procedures, it would usually
be irrelevant and unnecessary to state precisely who lit a burner
or soldered a joint, so we do not write:

The technician in our laboratory reacted A with B
My assistants and I connected the pin to the shank

We write, quite naturally:

A was reacted with B
The pin was connected to the shank
When the painted surfaces were examined last September
Data from the radar equipment were processed rapidly
Output from the reactor has been increased
Reed relays were fitted to the board, which was installed

Scientists are not unusual in using passive constructions in
this way. Four out of five English passive sentences have no
agent openly expressed[8]. We cannot object to this use of the
passive construction *in itself*. We *can* object to its *abuse* — to
its use almost to the exclusion of all other constructions. When
the passive is used as a rule, not as an exception to obtain a
particular effect, writing soon begins to seem forced and uncomfortable.

Though the proportions of active and passive constructions
used by individuals vary considerably, the active voice is the

predominant mode of expression for native English speakers. If it were not 'normal' for English statements to appear in the word order: *subject* (actor or agent) + *verb* (action or state) + *object* (person or thing affected), it would not be possible for us to gain particular effects by deliberately changing the word order or by changing the items used to fill the subject or object positions. To try to use only passive constructions is, therefore, bound to produce writing that seems artificial and forced; and as we have seen, it is likely to lead to distortion of meaning.

Since passive constructions are formed by adding parts of the verb *to be* (or sometimes *to get* or *to become*), plus *by* if the agent is expressed, continual use of passives commonly increases roundabout phrasing and length.   Increase in length is not inevitable. 'We calculated the yield' can become 'The yield was calculated'; but it frequently expands to 'Calculation of the yield was performed'.   Especially where an agent must be specified, extra words slip in.   'The ohmmeter measures the resistance' becomes 'The resistance is measured by the ohmmeter' or 'Measurement of the resistance is carried out by means of an ohmmeter'.   I shall comment in the next section (p.42) on the introduction of 'general-purpose verbs' such as *carried out*.   For the moment, let me just draw attention to the greater length of the first (mainly passive) versions in the following pairs:

The first test was carried out on the tower aerials in order to determine the vertical angle response.   A helicopter with a transmitter suspended beneath it flew to various points around the tower while radar equipment was used to track it, phases were measured by means of Measurit A and XY-5 performed calculations of the azimuth angle of the helicopter. The helicopter flew to six points spaced at equal intervals on a circle of radius about three miles centred on the tower. This procedure was carried out five times at heights of 1,000, 2,000, 3,000, 4,000 and 5,000 feet above ground level, the helicopter hovering for about 30 seconds at each point, and transmissions being produced for 15 seconds at 5·685 MHz and for 15 seconds at 8·975 MHz.

(In the first test to establish the vertical angle response of the tower aerials, a helicopter with a transmitter suspended beneath it flew to various points around the tower.   Radar equipment tracked the helicopter, Measurit A measured the phases, and XY-5 calculated the azimuth angle of the helicopter. The helicopter flew to six points equally spaced on a circle three miles in radius centred on the tower. It hovered for about 30 seconds at each point, transmitting for 15 seconds at 5·685 MHz and 15 seconds at 8·975 MHz. It made five circuits at 1,000, 2,000, 3,000, 4,000 and 5,000 feet above ground level.)

Mercaptans such as ethyl mercaptan, thiophenol and octyl mercaptan are converted by aqueous alkaline hydrogen peroxide to the corresponding disulphide. Yields are high. Acidified peroxide may be added so as to avoid cleavage of the disulphide by excess alkalinity and so that precipitation of the product may be assisted. In the case of branched chain high molecular weight mercaptans, some resistance to oxidation by hydrogen peroxide occurs.

(Aqueous alkaline hydrogen peroxide converts mercaptans such as ethyl mercaptan, thiophenol and octyl mercaptan to the corresponding disulphides. Yields are high. Acidified peroxide can be added to prevent cleavage of the disulphide by excess alkalinity and to assist precipitation of the product. Branched-chain high-molecular-weight mercaptans resist oxidation by hydrogen peroxide.)

The moiré effect is dependent upon the difference between the spacing of the lines and the pitch of the holes. The alignment of the brightness waves is not necessarily in parallel with the picture lines and any form may be assumed by the waves. The modulation depth of the waves is dependent upon the diameter of the luminance spot on the screen. No moiré effect will be experienced if the spot is large or if the lines are wide.

(The moiré effect depends on the difference between the spacing of the lines and the pitch of the holes. The brightness waves do not necessarily run parallel with the lines and may assume any form. The modulation depth of the waves depends on the diameter of the luminance spot on the screen. There will be no moiré effect if the spot is large or if the lines are wide.)

Note, too, that in improving the passive versions, it was not necessary to introduce personal constructions. *Active writing does not have to be personal.*

I suspect it is confusion about this that leads many teachers to enforce the 'rule' that scientific reports should be written in past-tense, passive-voice, impersonal constructions. Their intention is reasonable: they wish to develop in their students a detached, 'objective' outlook appropriate to scientific activity. But proper detachment does not require roundabout, passive thought and expression. Let us encourage objectivity by all means; but since passive-voice constructions are likely to distort meaning and increase ponderousness of expression, let us encourage incisive, active writing as the norm.

2.8 Personal and Impersonal Constructions

Excessive use of passive constructions creates artificiality and discomfort for readers. On top of this, the fact that passive constructions often do not specify clearly who was the actor or agent in an event frequently adds ambiguity and confusion. I misunderstood all the following extracts on first reading:

Smith and Robinson have determined the extent of thermal degradation by heating for standard times at 220°C and then calculating the weight loss during this time, together with the changes occurring in the infra-red spectrum of the polymers. The colour of the polymers has also been reported, as it gives some guide to the extent of the reaction. This stage of the investigation is nearing completion, and it is planned to examine the products of pyrolysis in the region of the exotherm using gas chromatography.

Green and White considered that the relative post-mortem stretching of the intestine is independent of the size of the animal, and it is especially rapid if the intestine is freed from the peritoneum.

After attempts had been made to measure the length of the pig intestine, it was realized that these measurements were of little value.

The following investigation was begun as a result of recommendations by John Brown in the report 'Measurements of Base-plate Profiles' (No. AB 123). In that report he noted that . . . In other work recorded in the above report it was found that the top edge suffered no appreciable dimensional change, even after several years' service. However, new base-plates often were outside specified tolerances. It was decided that the measurements should be made on a free-standing atmospheric rig, using air as the working fluid. Six plates were to be tested under varying...

In each of these, I missed a change of theme which was masked by use of impersonal constructions.

I discovered in discussion of the first extract, that 'This stage of the investigation is nearing completion' did not refer to Smith and Robinson's work but to the author's own work. If he had written 'In our investigation, we have almost completed... and we are planning' there would have been no confusion.

In the second example, it is not clear who attempted to measure the length of the pig intestine; it looks as if the second paragraph continues the account of Green and White's work. If the writer had begun his second paragraph with the words 'After we had attempted...we realized', we should have followed the transition from previous work to new work confidently and comfortably.

The third extract seemed at first to be entirely devoted to describing previous work by John Brown. In fact, the pen-ultimate sentence begins to describe decisions taken by the report writer in designing his own work. The transition could have been completely clear if he had written 'I decided that I would make my measurements'.

It would surely have been unreasonable to accuse these writers of presumptuous self-glorification for using personal constructions in these examples. There would have been no lack of humility or distortion of scientific fact; we should simply have been grateful to have the contrast between the earlier work and the new work clearly shown. The gain in clarity would have been considerable.

The need to preserve 'proper' scientific detachment and humility is often stressed by teachers who will not accept personal, active writing in scientific papers. They contend that readers of scientific papers are interested primarily in the scientific facts, not in who established them; the use of *I* or *we* is an unwarranted, immodest introduction of a specific agent into the account.

I readily agree with the need for clarity of perception and for balanced, detached comment. But I have argued already that there are points in scientific papers at which specification of an agent is essential to the clarity of the account. I have given three examples from introductions to papers. Here are more examples, this time from discussion sections:

— some study is being undertaken at the moment into the application of narrow aisle trucks in the warehouse
— consideration has been given to
— attempts have been made to prepare
— recommendations are made that
— the suggestion is made that
— it is recognized that these will be subject to change
— it is the intention that these methods be subjected to further refinement
— it is believed that eight radiators will be sufficient
— it has been remarked previously that
— it has been found by experiment that
— it has been established previously that
— it is claimed that
— it is thought that
— it is felt that
— it has been observed that

Whose work or ideas are being described here? Who is responsible for the opinions being expressed? Who is making or has made the claims or remarks? Who thinks or has thought, who feels or has felt or has made the observations? Readers do not know. 'It...that' constructions in particular often create vagueness or a sense of hesitancy or evasion. Impersonal writing of this type should be discouraged, not enforced.

I contend that, in the contexts in which these extracts appeared, the clarity of the messages would have been increased considerably if the author(s) had written:

— we are currently studying the use of narrow aisle trucks in the warehouse
— we have considered
— we have tried to prepare
— I suggest that
— I recognize that these will be subject to change
— we intend to refine these methods further
— I believe that eight radiators will be sufficient
— I have remarked previously that

— in previous experiments, we have found
— we have established previously that
— we claim/believe
— I think
— I feel
— we have observed/noticed

My over-riding concern in all this advice is for clarity and accuracy of communication. Automatic adoption of 'It...that' constructions frequently reduces both clarity and accuracy. Though such constructions are legitimate and natural features of English, they should be used sparingly and with precise intention.

Regrettably, writers get locked into an 'It...that' frame of mind, and begin to use it when it is totally unnecessary.

It has been observed by Green and White that
(Green and White have observed)

it was considered by Muguruma (1961) that
(Muguruma (1961) considered that)

in an earlier paper (Smith 1961) it was shown that
(in an earlier paper (Smith 1961) I showed)
(in my 1961 paper, I showed)

Equally regrettably, 'It...that' constructions are sometimes used deliberately to suggest wider support or acceptance for a statement than is warranted. Dr I. J. Good has alerted scientific readers to the real meaning of many such impersonal constructions[9].

It is clear that much additional work will be required before a complete understanding
(= I don't understand it)
It has long been known that
(= I haven't bothered to look up the original reference)

I have no quarrel with genuine attempts to present facts with detachment and diplomacy; but I have no time for the spurious 'objectivity' that produce "'ly circumlocutions such as:

it was speculated that
(I thought, I speculated, I guessed)
it is highlighted that
(we would stress that)

it is strongly considered that
(I firmly believe/consider)
it was felt subjectively that smooth operation was not likely
(I thought smooth operation was unlikely)
it is suggested by the author that
(I suggest that)
the results lead the present authors to believe that
(the results lead us to believe that)
before the experiment, the author was very uncertain how to
(before the experiment, I was very uncertain how to)
it is the author's preference to install
(we prefer to install)
the author was glad to avail himself of the opportunity to
make vibration measurements
(I was pleased to take the opportunity to)
it is the writer's opinion that
(in my opinion)
no details are available to the present writer on the specification of
(I have no details of the specification of)
in the present author's experience
(in my experience)
the author's wife, Mrs John Smith, typed the thesis
(my wife typed the thesis)

Once again, I am pointing out that the 'rule' is bad not because it advocates the use of a particular form but because it advocates the use of that form almost to the exclusion of all others. Impersonal constructions have a proper and useful place in our natural use of English. But they are not automatically interchangeable with personal constructions.

Consider the following examples: the personal and impersonal statements do not have identical meaning:

I do not accept the idea
One cannot accept the idea
The idea cannot be accepted

I suspect that
It is possible that
A suspicion may arise that

We do not understand how the chlorine reacts with
It cannot be understood how the chlorine reacts with

    gave one of the strongest wiremarks we have ever seen
    gave one of the strongest wiremarks ever experienced

    and we have information that
    and there is information that

The last three pairs were changes from personal to impersonal constructions made by senior staff 'editing' young scientists' texts. The editing was harmful. Writers must not be required artificially to force all their thoughts into an impersonal mould: there is real danger that they will distort their intended meaning.

Most scientists would be horrified at the thought of unrecognized inaccuracy in their experimental work; yet an astonishing number cheerfully ignore vaguenesses and approximations of meaning that creep in as they straitjacket their ideas into stereotyped patterns of expression.

The habitual use of impersonal constructions has an insidious inflating effect. Writers who will not (or feel they *must* not) write:

    we sampled the ions from the plasma by
    I removed the coating with alcohol
    we did not inspect the burners regularly

can write in simple passive form:

    the ions from the plasma were sampled by
    the coating was removed with alcohol
    the burners were not inspected regularly

But time and again they yield to temptation to take a further step and expand these statements to:

    ion sampling from the plasma was achieved by
    removal of the coating was effected by the application of alcohol
    regular inspections of the burners were not carried out

In taking this extra step they not only change the construction from personal to impersonal and from active to passive: they also introduce colourless 'general-purpose' verbs carrying abstract nouns. They no longer *sample, remove* and *inspect*; they *achieve, effect* and *carry out*. And the focus of their statements is no longer on *ions*, the *coating* or the *burners*; it moves to the vaguer abstractions *sampling, removal* and *inspections*.

Undoubtedly, there are many occasions when we want to focus on *sampling* rather than on *ions,* or on *inspections* rather than on *burners.* In such circumstances, as I have argued earlier, the use of a passive construction is a proper and valuable way of moving the emphasis of a sentence. I repeat that I am not condemning *all* use of passive constructions: I am simply anxious to show the pernicious inflating effect of using impersonal passive constructions with 'general-purpose' verbs as the rule rather than as the exception.

Much of the ponderous monotony of scientific writing is contributed by this practice. Just consider how much time scientists apparently spend *performing*:

measurements of reflection and transmission are performed
(reflection and transmission are measured)

the integration of X with Y may be performed
(X may be integrated with Y)

an experiment was performed on a small scale to ascertain the influence of X, which resulted in
(in a small-scale experiment to ascertain the influence of X)

our facilities were inadequate to perform the experiment
(our facilities were inadequate for the experiment)

our present measurement of surface area is performed by
(at present we measure surface area by)

calculations of the yield were performed which revealed
(we calculated the yield and found)

a second trial was performed
(in a second trial, we)

and *conducting*:

tests have been conducted on the pavements
(the pavements have been tested)
(we have tested the pavements)

a test was conducted to discover the value of X
(in a test, the value of X was)

another study of the roof was conducted twelve months later
(the roof was studied again twelve months later)
(we studied the roof again twelve months later)

trials were not initially conducted on operational aircraft
(at first, there were no trials on operational aircraft)

and having *experiences*:

unusual shrinkage of the diaphragm plaque was experienced
(the diaphragm plaque shrank unusually)

a high number of breakdowns was experienced in March
(there was a high number of breakdowns in March)

quantities in excess of one litre were experienced
(more than one litre was produced)

greater success was experienced on the plant in reducing the
free acid concentration than was experienced in the laboratory experiments
(on the plant, the free acid concentration was reduced more
successfully than in the laboratory experiments)
(on the plant we reduced the free acid concentration more
successfully than in the laboratory experiments)

They are great *carriers* too:

volume control of the individual speakers is carried out by a
switch
(a switch controls the volume of the individual speakers)

regular inspections of the burners are not carried out
(the burners are not inspected regularly)

periodic cleaning and lubricating of the fan filters should be
carried out every six months
(the fan filters should be cleaned and lubricated every six
months)

to date, most work has been carried out on sheets of
(to date, most work has been on sheets of)

daily measurements of levels are carried out
(levels are measured daily)

installation of the units will be carried out by works personnel
(works personnel will install the units)

a final test was carried out consisting of running the rig
continuously
(in a final test, the rig was run continuously)
(in a final test, we ran the rig continuously)

refinement of this plan was carried out
(this plan was refined)
(we refined this plan)

load-cycle tests were then carried out
(load cycles were then tested)
(we then tested load cycles)

measurements of electron and ion densities in argon afterglows were carried out
(electron and ion densities in argon afterglows were measured)
(we measured electron and ion densities in argon afterglows)

and they *accomplish* great things:

vacuum sealing of the waveguide itself was accomplished
using quartz windows
(the waveguide itself was vacuum-sealed using)

heat-treatment of the film was then accomplished
(the film was then heat-treated)
(we then heat-treated the film)

Expressions such as *was performed, were conducted, were
experienced, were carried out, was achieved, was shown, were
effected, were observed, were accomplished, resulted* and *occurred* are desperately overworked in scientific writing because
scientists are reluctant to write directly and personally. Notice
that it was possible to remove most of the roundabout expressions without introducing a personal construction; but would
introduction of personal constructions necessarily have been
undesirable? Is it true that occasional introduction of personal
constructions *always* strikes scientific readers as obtrusive or
immodest?

The reactions of the respondents to my surveys deny this.
The majorities *preferred* texts that used a judicious mixture of
personal and impersonal constructions (pp.125-7). There was
some confused thinking:

I would have given version H [using personal pronouns]
first preference had the 'we's' been eliminated, e.g. by 'it was
felt', not 'we thought'...

Version H [using personal pronouns] would have been easily
first if written in the third person...

Many who were reluctant to vote in favour of texts that included personal pronouns acknowledged that those texts were very clear.   Many admitted, too, that their reaction against those texts was probably due to brainwashing during their education, not to any rational objection (pp.128-30).

Let me stress that in advocating the use in writing of the personal, active phrasing of serious discussion, I am not advocating the use of the casual, inexplicit shorthand of person-to-person chitchat.   I am not advocating that the account should be *over*-personalized, sensationalized or trivialized.   *Over*-indulgence in personal constructions is just as distorting as *non*-indulgence: to sound self-congratulatory is as bad as to pretend you were not involved at all.   And I am not suggesting that the scientific content should be distorted in any way.   I *am* advocating that writers should write naturally and economically, without affectation of a special 'scientific' style.   They should come as close as possible to the natural mixture of constructions and the natural balance and rhythms of comfortable everyday speech.

My advice has sometimes been misrepresented as encouragement to writers to claim credit for work they did not do themselves.   That might be true if I advocated that *all* impersonal, passive constructions should be replaced by *I* or *we*.   I make no such suggestion.   If much work has been done on a project by someone other than the writer or the writer's group, then it is entirely appropriate to write 'much work has been done' or 'in many experimental programmes it has been found that', with or without attribution to a particular agent.   I advocate simply that writers should make statements active wherever possible.   For example 'installation of a noise barrier has been carried out by Smith and Jones' should be converted to 'Smith and Jones have installed a noise barrier'.   In particular, where writers have been personally involved in some work, they should not use cumbersome passive constructions simply to avoid saying *I* or *we*.   They should make clear distinctions between their own observations, beliefs and deductions and those of other writers.

Once again, I am suggesting that it is the blanket operation of the 'rule' that is harmful.   Writers should be taught that the patterns of expression naturally required will probably vary in different parts of their reports and papers.   Introductory sections will be the most varied in tense, voice and person.   Descriptions of procedures will be mainly past tense and impersonal, but should be kept as active as possible.   Conclusions and recommendations will vary in tense and voice. Since they convey the writer's conclusions and recommendations, they can quite properly be personal; but both will often be most effectively expressed as lists:

Conclusions
1. A policy of part-life-rework for ABC engines is essential.
2. The engine bay at Someplace works could cope with the work involved in operating such a policy.
3. A part-life-rework policy would necessitate the issue of revised modification standards and service operating procedures.

Recommendations
1. that the guard leads be screened in plastic (para 5.3);
2. that the gear ratio to the shaver be reduced to 10:1 (para 6.4);
3. that the measuring head be fitted with a dust shield (para 7.2).

I have been told that it is essential to train students — especially research students — to write impersonally and passively because most editors of professional journals will not accept articles that contain personal constructions. That simply is not true. In a walk round the Periodicals Library at UWIST, I took the current copies of 81 scientific journals at random from the shelves. I found articles using *I, we* or *our* in 74 of them:

Acta Crystallographica
Advances in Physics
American Journal of Physics
Biochemical Journal
Biological Conservation
British Journal of Industrial Medicine
British Journal of Ophthalmology
British Journal of Nutrition
British Journal of Pharmacology
British Medical Bulletin
British Medical Journal
Bulletin of Entomological Research
Chartered Municipal Engineer
Chemical Engineer
Chemical Engineering Science

Computer-Aided Design
Computer Journal
Contemporary Physics
Cryogenics
Econometrica
Electrical Review
Electronics and Power
European Polymer Journal
Experimental Eye Research
Faraday Discussions of the Chemical Society
Faraday Symposia of the Chemical Society
Hydrographic Journal
IEEE Transactions
Inorganic Chemistry
International Journal of Electronics
International Journal of Engineering Science
International Journal of Mechanical Sciences
International Journal for Numerical Methods in Engineering
Journal of the American Chemical Society
Journal of Applied Ecology
Journal of the Chemical Society, Faraday Transactions
Journal of the Chemical Society, Perkins Transactions
Journal of Ecology
Journal of Environmental Management
Journal of Fluid Mechanics
Journal of the Geological Society
Journal of the Institute of Mathematics and its Applications
Journal of Materials Science
Journal of the Mechanics and Physics of Solids
Journal of Navigation
Journal of Pharmacy and Pharmacology
Journal of Physics
Journal of Physiology
Journal of the Royal Statistical Society
Journal of the Science of Food and Agriculture
Lancet
Manufacturing Technology
Microelectronics Journal
Mikrochimica Acta
Nature
Ocean Management
Operational Research Quarterly

Philosophical Magazine
Physica Status Solidi
Polymer
Polymer Engineering and Science
Proceedings of the American Society of Civil Engineers
Proceedings of the London Mathematical Society
Proceedings of the Royal Society of London
Progress in Materials Science
Psychopharmacology
Quarterly Journal of Mechanics and Applied Mathematics
Quarterly Journal of the Royal Meteorological Society
R & D Management
Radio and Electronic Engineer
Spectrochimica Acta
Transactions of the American Society of Mechanical Engineers
Transactions of the Institution of Chemical Engineers
Water Resources Research

The remaining seven seemed entirely impersonal:

Journal of the Institution of Water Scientists
Journal of Mechanical Engineering Science
Machine Tool Design and Research
Powder Technology
Proceedings of the Institution of Civil Engineers
Strain
Water Pollution Control

Two notes must be added.

1. The omission of a particular journal from these lists is not a slight on that journal: it implies simply that I did not get round to it in my random selection. I gave up browsing after two hours, feeling that my point was amply supported.

2. I do not know whether the seven journals that seemed impersonal have a deliberate policy to be so. They, too, may be prepared to accept good personal writing if they receive it.

The way in which we permit ourselves to select language is of outstanding significance, because it goes a long way towards establishing our over-all 'mental set', our customary habit of attack on our daily tasks of thought and expression. Are we to think as directly, positively and flexibly as possible, or are we to approach every statement in a stereotyped, roundabout,

non-committal, self-protective way? Certainly, there are times
when we have to write diplomatically. Certainly, there are
many contexts and topics — for example, descriptions of
machines or reactions — which do not naturally call up personal
constructions. But in most scientific papers and reports, we are
concerned to be as direct, explicit and economical as we can.
In those circumstances, we should write in a natural, comfortable mixture of personal and impersonal constructions, using
active verbs as our main mode of expression and interweaving
passive verbs skilfully to change the balance and emphasis of
what we want to say.

2.9 Punctuation

This is not the place for a full discussion of modern conventions
of punctuation. Readers who would like general guidance
should look at the short sections on punctuation in *Modern
English Usage*[10] or *The Complete Plain Words*[11]. For more
extensive discussion, I would recommend the short books *Mind
The Stop*[12], by G. V. Carey and *You Have a Point There*[13],
by Eric Partridge.

Suffice it to say here that both too little and too much
punctuation are distracting. A constant irritation that I find in
scientific writing is the omission of hyphens. Sometimes the
collocation of words without hyphens makes such nonsense
that readers have no difficulty in detecting the intended
meaning:

> made counts of the bacterial loads in the air. He observed
> that bacteria carrying dust particles decreased in concentra-
> tion as the humidity decreased
> (bacteria-carrying)

> can be present in boron containing steels
> (boron-containing)

But readers should not have to work out the correct relations
between words because writers have failed to signal them
clearly. It should not be necessary to read the following
examples twice in order to be sure what they mean:

> Typical chromatograms for pure morpholine solutions and
> for morpholine containing effluent liquors are shown
> (morpholine-containing)

will give two dimensional plots of thickness contours
(two-dimensional)

prefer to use metal cutting machine tools
(metal-cutting)

before one can discuss plans to cut down noise one must
isolate the noise creating factors
(noise-creating)

taking off gases from the esterifiers
(off-gases)

Absence of commas, especially to mark off words and clauses
at the beginnings of statements, is another regular source of
ambiguity and irritation.

Frequently adjusted totals need to be scrutinized
(Frequently, adjusted)

Should any burner fail to ignite its respective section will
revert to 'purge' and in this way the
(fail to ignite, its respective)

As the truck passed the vibrations were noticeable to anyone
standing on the bridge
(the truck passed, its vibrations)

As the machine develops the forms we use to record data
from past projects will be amended
(the machine develops, the forms)

In the event, therefore, of failure of any of these components
to operate the respective motor will not start until the fault
has been rectified
(components to operate, the respective motor)

Punctuation marks are essential elements of the code on
which written communication is based. We are not at liberty to
leave them out or to sprinkle them at random over our work.
Here is one example of too little punctuation, and one example
of too much:

Normally journal bearings are white metal lined sleeve type
and the thrust bearings are the tilting pad type, which also
provide axial location of the rotors. All bearings are oil
pressure lubricated. In lighter duty machines ball and roller
bearings may be employed to reduce mechanical losses.

The wires connected between this voltage source and the sound unit are the means, by which, the energy is carried between the two units.   Thus, with the power source connected up, the driving force is taken to the sound unit, but still, no sound can be heard.   No sound will, in fact, be heard, until the push-button is depressed.

As we read the first example, we have to supply several punctuation marks.   This means that we move jerkily along the lines of print, working out what logical groups and nuances of tone the writer wants us to recognize.   In the second example, too, our progress is jerky, but this time because the writer interrupts us too frequently with commas.   The following improved versions allow us to move along smoothly without constant checking back to see if we are putting the right constructions on the words:

Normally, journal bearings are white-metal-lined, sleeve type and the thrust bearings are the tilting-pad type, which also provide axial location of the rotors.   All bearings are oil-pressure lubricated.   In lighter-duty machines, ball- and roller-bearings may be employed to reduce mechanical losses.

The wires connected between this voltage source and the sound unit are the means by which the energy is carried between the two units.   Thus, with the power source connected up, the driving force is taken to the sound unit, but still no sound can be heard.   No sound will, in fact, be heard until the push-button is depressed.

The full impact of punctuation and paragraphing is difficult to demonstrate in short passages such as were used in the surveys.   Also, I decided that it would be unfair to load the scales heavily against some versions by deliberate mis-punctuation, which would have made comprehension difficult.   My aim was to stop any of the versions seeming obviously 'wrong', confused or unacceptable because of glaring faults of organization, grammar, punctuation or spelling.

Nevertheless, the effects of paragraphing and of varying sentence lengths were obvious; I hope readers will be interested by the effects achieved in the six versions in each survey by varying punctuation.

G. V. Carey's philosophy, expressed in *Mind the Stop*[14], is a good one to have in mind:

Stops should be used as sparingly as sense will permit: but in so far as they are needed for an immediate grasp of the sense or for the avoidance of any possible ambiguity, or occasionally to relieve a very lengthy passage, they should be used as freely as need be. The best punctuation is that of which the reader is least conscious; for when punctuation, or the lack of it, obtrudes itself, it is usually because it offends.

References

1. Tinker M. A., *Legibility of Print*, Iowa State University Press, 1963.
2. Turnbull A. T. and Baird R. N., *The Graphics of Communication: typography, lay-out and design*, Holt, Rinehart and Winston, 1968.
3. Macdonald-Ross M. and Smith E., *Graphics in Text: a bibliograph*, IET Monograph No. 6, Open University, 1977.
4. Fowler H. W., *Modern English Usage* (second edition, revised by Sir Ernest Gowers), Oxford University Press, 1965, p.461.
5. Gunning R., *More Effective Writing in Business and Industry*, Farnsworth Publishing Inc., 1962, pp.5-29.
6. Gowers E., *The Complete Plain Words*, Pelican Books, 1962, p.136.
7. Pickering G., 'Language, the Lost Tool of Learning in Medicine and Science', *The Lancet*, 15 July 1961, p.116.
8. Quirk R., Greenbaum S., Leech G. and Svartvik J., *A Grammar of Contemporary English*, Longman, 1972, p.807.
9. Good I. J., *The Scientist Speculates*, Heinemann, 1962, pp.52-53.
10. Fowler H. W., *Modern English Usage* (second edition, revised by Sir Ernest Gowers), Oxford University Press, 1965, pp.587-592.
11. Gowers E., *The Complete Plain Words*, Pelican Books, 1962, pp.236-260.
12. Carey G. V., *Mind the Stop*, Cambridge University Press, 1960.
13. Partridge E., *You Have a Point There*, Hamish Hamilton, 1964.
14. Carey G. V., *Mind the Stop*, Cambridge University Press, 1960, p.12.

3 Style for Instructions and Procedures

Operating instructions are written so that a process, a procedure, a plant or a piece of equipment can be operated unhesitatingly, accurately and economically. The instructions must therefore be:

accurate: the exact meaning of each instruction must be clear;

comprehensible: each instruction must be manageable, not too involved or overloaded with information, not blurred by explanation and description;

adequate: there must be enough information to permit operation, or at least there must be exit branches specifying where additional information may be found;

complete: all feedback loops and branches must be specified;

in sequence: the steps must be in proper order;

safe: warnings must be prominent and well placed;

acceptable: the tone must help gain the response desired; it must not produce resentment or hostility.

To meet these requirements, it is necessary to go beyond consideration of style. We must consider the selection, arrangement and physical presentation of information; we must have in mind constantly the frames of reference of the receivers, their levels of expertise, their past training (if any), and the contexts in which the instructions have to be used. A full discussion of tactics for organization and layout of instructions is beyond our scope here; but consideration of the following example will make plain how features of style are inter-related with features of selection and organization:

Tubular Unit Operating Instructions

1. . . . . .
2. . . . . .
3. . . . . .
4. . . . . .
5. The furnace interior should be inspected every four hours for insufficient burner operation. Look for ragged flame, smoke or 'points of flame'.
6. Check the stripping column bottoms level controller AB12. Raise the set-point on AB12 to ...X.... Level pen should follow. Float freedom can be tested by means of the rod in the base.
7. Twice a shift check the whole system for process leaks. Take particular care in the furnace and pre-heater area.

Does the first sentence in Instruction 5 give advice or an instruction? 'The furnace interior *should* be inspected' seems to leave an option open.

We are told to look for 'ragged flame, smoke or "points of flame"'; but we are not told what to do if we find some!

We are told to check the stripping column bottoms level controller AB12. What does this mean we have to do — just look at it?

Presumably, raising the set-point to ...X... will cause no difficulty. But does it matter if level pen does *not* follow? And what should we do if it does not?

The final sentence in Instruction 6 is not an instruction: it is a comment. What relevance does it have to the instructions? When should float freedom be tested? Should the float be free? What should we do if it is not?

Twice in each shift we are to check the whole system for process leaks. If we are on an eight-hour shift beginning at 8 am and it takes 20 minutes to walk round the system, is it satisfactory for us to walk round once from 0800 to 0820 and again from 0830 to 0850? And what are we to do if we find any leaks? Merely write a note in the log book? Shut down the plant? Run?

We are told to take particular care in the furnace and pre-heater area. Why? Is it slippery in there?

Plainly, the adequacy of these instructions would depend greatly on the expertise and previous training of the users. But even if the users had been given a detailed training course, there should have been some cross-references in these instructions to relevant parts of training manuals or other explanatory documents; and many of the ambiguities could have been removed by more careful choice of language and style.

The danger of a vague word like *should* stands out obviously. Each use of it in any instructions should be scrutinized closely, as it may reduce to a comment an item intended as an instruction. If it is necessary for a reactor to be warmed for two hours before use, to write 'the reactor should be warmed' will not be adequate.   You will have no grounds for complaint if the warming is omitted, as you have used an optional form: you must write either 'the reactor must be warmed' or 'warm the reactor'.

The word *will* can similarly reduce an instruction to an observation. 'The filter will be changed every six hours' simply states a fact: no-one is instructed to change the filter. To make this an instruction, you must use expressions such as 'the filter is to be changed', 'the filter must be changed' or 'change the filter', and you must specify who is to change it.

Another vague word is *check*.   Again, you will have inadequate grounds for complaint if nothing happens in response to the instruction 'check that the drain valve to the top of the column is closed'. Is the operator simply to report the state in which he finds the valve? If you want him to close it, you must write 'ensure that the drain valve to the top of the column is closed'.

Other dangerous words are:

select: 'select first gear': does this mean there is a choice of which gear is to be used first?

locate: 'locate the retaining knob on the baseplate': does this mean *find* or *connect*?

replace: 'replace the pin that is damaged': does this mean *put it back* or *fit a new one*?

To be safe, it is necessary to define service and process activities clearly. Probably, the best place to do this is in a training programme, with clear definitions written in a training manual; but if necessary, it should be done at the start of each document, as is attempted in this example from a rescue harness servicing procedure[1] :

SERVICING NOTES

1. *Glossary*	The servicing operations detailed in this bay servicing schedule have the meaning given in the Concise Oxford Dictionary except for the following:
a. *Inspect*	Review the work carried out by tradesmen to ensure it has been performed satisfactorily.
b. *Check*	Make a comparison of a measurement of time, pressure, temperature, resistance, dimension or other quantity with a known figure for that measurement.
c. *Test*	Ascertain, by using the appropriate test equipment, that a component or system functions correctly.
d. *Examine*	Carry out a survey of the condition of an item. For example, the condition of an item can be impaired by one or more of the following:

    (1) Insecurity of attachment.
    (2) Cracks or fractures.
    (3) Corrosion, contamination or deterioration.
    (4) Distortion.
    (5) Loose or missing rivets.
    (6) Chafing, fraying, scoring or wear.
    (7) Faulty or broken locking devices.
    (8) Loose clips, or packing, obstruction of, or leaks from pipelines.
    (9) Discoloration due to overheating or leaking of fluids.

e. *Operate*	Ensure an item or system functions correctly, as far as can be ascertained without the use of test equipment or reference to measurements.
f. *Replenish*	Refill tank, bottle, or other container, to a predetermined level, pressure or quantity, and where necessary:

    (1) Remove caps, or covers from filler orifices and/or drains.
    (2) Clear orifices.
    (3) Fill container as directed in item operation.

	(4)	Ensure drains are free from obstruction.
	(5)	Ensure gaskets and caps or covers are free from damage.
	(6)	Refit caps or covers.
	(7)	Fit locking devices as necessary.

g. *Fit*	Correctly attach one item to another.
h. *Refit*	Fit an item which has previously been removed.
j. *Replace*	Remove an item and fit new or serviced item.
k. *Disconnect*	Uncouple or detach cables, pipelines or controls.
l. *Reconnect*	Reverse of sub-para k.

Each word or phrase must be considered in relation to the readers' background knowledge. Will they understand expressions such as:

> Slacken the cheesehead retaining screw
> Remove the camerated segment
> Examine the wheel disc for signs of fretting
> Slide the guideplate right home
> Pull smartly upwards

The first three examples use specialist jargon; will all readers understand it? If so, by all means use it. The last two examples make use of images familiar to native speakers of English: but will they be comprehensible to readers for whom English is a second language?

Constant attention to the readers' frames of reference is vital. Do they know what is meant by 'connect up the supply tank and the liquid gas vaporizer *in the usual way*' (my italics)? Do they have in mind the same ranges as you have when they read your instruction to 'ensure there is *little or no* pressure in the reactor' (my italics) or 'set the pressure in the main to *about* 40 p.s.i.' (my italics)?

And will it be clear who is to act in each instruction? Clear indication of the agent, if it is necessary to specify one, will depend on two things: the format of the complete set of instructions and the style.

If you specify at the beginning of a set of instructions the readers for whom you are writing, you can write freely in impersonal style. For example:

Instructions for Bagging Machine Operators
The filter is to be changed every six hours
The filter must be changed every six hours
Change the filter every six hours

But if you have not specified whose instructions these are, you may find the change not made 'because it didn't say it was my job to do it!'
Occasionally it is desirable to specify an exact agent:

The supervisor/Mr Brown is to change the filter
The supervisor/Mr Brown must change the filter

In these circumstances, too, you may find that the change has not been made 'because it says it's Brown's job, not mine!' This risk must be guarded against by ensuring that your work organization has specified who is to act in Brown's absence.

We are now beginning to consider shades of *tone* in operating instructions — the influence style can have in getting not only accurate but also willing operation of processes or equipment. The impersonal and imperative styles I have mentioned so far will normally produce instructions that are both clear and acceptable in tone. But there is frequently a difference in response to second-person styles of instruction writing. If you write: 'You are to change the filter every six hours' or 'You must change the filter every six hours' many readers will feel that the instructions are too sharply domineering and autocratic. Their reaction is against the use in *written* instructions of the form 'You'. Interestingly, the 'You' form of instruction when *spoken* is normally quite acceptable; but when frozen in writing, the 'You' form takes on a more pointed tone and will normally be resented by readers, particularly when it is coupled with the infinitive form 'are to'. The instruction 'You must change the filter' will be just about acceptable, as it seems simply to state what must be done as a result of the natural outcome of events. 'You are to change the filter' carries an autocratic tone: you are to change it *because I say so*, and that will be resented.

For clarity and acceptability, instructions are best written as a sequence of steps, each beginning with the imperative form of a functional verb (raise, lift, turn, switch, ensure, connect). Occasionally, siting or conditional phrases can appear before the verb:

At panel B, switch...
If the reactor is partly full, connect...

Occasionally, for variety or emphasis, it may be useful to turn to the impersonal style or to specify an agent or to turn to the *must* or infinitive forms; but a crisp sequence of short, imperative instructions will normally produce best results.

Two final points can be made with one more example:

Flanged Valves: Test Procedure

1. Ensure cleanliness as described under 'Preparation' paragraph.
2. If a blow down ring is fitted, remove the blow down ring lock screw and then turn the tooth towards the right (counter-clockwise viewed from above) thus raising the blow down ring until it meets the disc holder, then lower by turning the tooth towards the left (clockwise viewed from above) to half the number of teeth shown in either table A or B, (depending on service of valve), but not less than two teeth. This may be done by reaching in through the locking screw hole with a screwdriver. After the blow down ring is relocated, replace the lock screw and tighten; make sure the lock screw rests BETWEEN the teeth so as not to distort the ring. Side pressure on the blow down ring can cause leakage. The blow down ring should have some play after the screw has been tightened, but should not be free to rotate.
   NEVER RAISE THE RING WHILE PRESSURE IS ON THE VALVE, AS THE VALVE MAY 'POP' AND COULD CAUSE INJURY.

The first point illustrated here is the clumsiness of writing that attempts to amalgamate several instructions into a single long paragraph. How much clearer these instructions would have been as a sequence of short, imperative instructions!

The second, *very* important point, is the position of the warning. After telling us at the top of the paragraph to raise the blow-down ring, the writer warns us at the end that to do so while pressure is on the valve may be dangerous. Too late! I am constantly astonished by the frequency with which warnings are given *after* dangerous steps in instructions. The correct place for warnings is *before* the dangerous step. And warnings should be clearly separated, not buried in the text, as in:

The dilution of sulphuric acid should preferably be carried out in clean earthen-ware or lead lined tanks which should be about half filled with pure water, and a suitable quantity of concentrated acid poured slowly into it, the mixture being stirred all the time with a clean piece of wood. Concentrated acid is much heavier than water, and will tend to settle at the bottom. On no account should the water be added to concentrated acid, as the water would remain on the surface and the intense heat generated would cause it to boil and splash the corrosive liquid about. The effect might be so violent as to be almost explosive.

Reference

1. Royal Air Force, Central Servicing Development Establishment. Crown Copyright, reproduced by permission of the Controller of Her Majesty's Stationery Office.

4 Style for Expository Writing and Specifications

4.1 Descriptive Handbooks and Technical Documentation

Most handbooks and other forms of technical documentation consist of description, explanation and instructions. They contain little theoretical analysis, evaluation, argument or expression of opinion. The most suitable style for most handbooks and technical documents is therefore likely to be impersonal. There will be little natural call for the personal constructions suited to debate and evaluation.

But to say the style should be impersonal is not to say it should be ponderous and dull. To write impersonally does not necessarily require use of a passive, roundabout style:

NOT: In so far as the carriage of current is concerned this is achieved through the medium of bus-bars.

NOT: The current is carried by means of bus-bars.

BUT: Bus-bars carry the current.

Expository writing can be clear, lean and vigorous. For example, compare and contrast the two versions of a description below. The 'better' version is lucid and readable. Its sentences are of varied length but on average they are short and not complex. Verbs are mainly active and the vocabulary is 'plain' but accurate. It illustrates, I hope, the cumulative benefits of writing directly and actively:

NOT: The extrusion or capillary rheometer, which is capable of recording a high shear-rate range, and may be utilized to obtain a measure of normal stress, is generally considered to be the most suitable apparatus for the determination of the characteristic flow behaviour of unvulcanized silicone rubbers. In this apparatus, extrusion of the material under consideration is carried out through a cylindrical die. The apparatus consists essentially of a cylinder in which the material is placed and a ram by which the rubber is forced along the cylinder, being extruded through a die at the end of the cylinder. The apparatus is attached to a tensometer and measurements of the force required to force the material through a die at different rates may be determined. A curve relating the volume extrusion rate and the back pressure may be produced without difficulty by the operator

BETTER: The extrusion or capillary rheometer is the best for finding out how unvulcanized silicone rubbers flow. Its shear-rate range is high, and it can be used to measure normal stress.

In this rheometer, a ram forces the rubber through a cylinder and out through a die at the end of the cylinder. A tensometer attached to the apparatus measures the force required to push the rubber through a die at different rates. The operator can easily produce a curve relating the volume extrusion rate and the back pressure.

A particular difficulty in descriptions and discussions is to keep clear in the reader's mind the difference between features or possibilities that are hypothetical or possible and those that do exist in the equipment, process or system.

The clarity with which this distinction is made depends greatly on careful control of the sequence of verb forms. If a statement begins with a condition, the verbs that follow must be mainly conditional forms. The sequence for discussing at a hypothetical level a fact or condition that *could* or *might* exist now or in the future should be:

If X *were* true (at the moment, or in the future), Y *could/would/might* happen, and Z *could/would/might* be possible.

The sequence for stating that, though we have no supporting evidence, we are accepting for the moment that a fact or condition *does* exist should be:

If X *is* true (now), Y *can/will/may/could* happen and Z *is/can/will/may be/could be* possible.

The sequence for discussing one or more possibilities *in the future* should be:

> If X *is/becomes* true (at some time in the future), Y *may/ will/could* happen and Z *is/may be/will be/could be* possible.

The sequence for expressing a fact or condition that *could have/ may have* existed at a specified time in the past, though we do not know whether or not it *did* exist should be:

> If X *was* true (at that time), Y *may have/could have happened/ been happening* and Z *may have been/could have been* possible.

The sequence for expressing a fact or condition that could have been true, but was not, at a specified time in the past should be:

> If X *had been* true (at that time), Y *would/could/might have* happened and Z *would/could/might have been* possible.

Unfortunately, writers all-too-frequently confuse what *is* true with what *might* or *could* or *should be* or *will be* true:

> In order to carry out a major examination on any one transformer the following procedure should be followed. The 11,000 volt 250 MVA OCB in the Engine House should be isolated and drawn out. A notice should be placed on the switch to give warning of work being carried out on the system. The 33,000 volt 1,000 MVA OCB is then tripped, either from the Engine House or from the outdoor substation. The control panel door on the 1,000 MVA OCB was then opened and the Control Key removed, this would then trip the 1,000 MVA OCB if the above methods were not followed. The Control Key is then used to unlock the mechanism on the pole isolator. After isolation the mechanism is again locked using the Control Key. The system should then be earthed to drain off any remaining charge.

As a reader, I was unable (even though I had the help of surrounding information) to make out which meaning the writer wanted me to take from this part of his report:

> The simple index might not have been valid in this case because the lateral spacings are also changed and this could affect the magnitude/sharpness relationship explained in the notes.

Was he talking specifically about the implications of one case he had just described:

The simple index possibly was not valid in this case because the lateral spacings also were changed and this would have affected the magnitude/sharpness relationship explained in the notes.

Or was he pausing to reflect on the general implications of his findings:

The simple index may not be valid in cases such as this because the lateral spacings also are changed and this could affect the magnitude/sharpness relationship explained in the notes.

One difference in tactics for descriptive writing is that occasional use of a terser, smaller-step style of organization and expression can be helpful to readers. Such a style would seem 'bitty' and irritating if it were used for long stretches of writing: it would certainly be unsatisfactory as a style for sustaining an argument or discussion in a report. But readers who are following an account of a complex process, especially those who are trying to *learn*, not just to understand, the information being presented, often find it helpful if *parts* of the text are presented in small, easily digestible steps.

Compare the following passages. The first version is in cumbersome 'traditional', passive, involved style. The second version is more direct and readable. The third version demonstrates how this type of information can be stepped and emphasized for particular purposes:

Completion of Supertron triggering is followed by commencement of anode run down and coupling back of the negative going waveform to the grid through V54A and a timing capacitor selected by Sw9/5, causing a 'Miller' type run down, at a speed controlled by the selected timing capacitor and RV51 setting. This run down, 'caught' at 106 volts by V52B, preventing possible jitter when the valve 'bottoms', is followed by regenerative stable state resumption by the Supertron, causing a sudden screen potential fall differentiated by C185 and R378, and triggering of the time base.

After the Supertron has been triggered, the anode begins to run down and the negative-going waveform is coupled back to the grid through V54A and a timing capacitor (selected by Sw9/5). This gives a 'Miller' type run-down. The speed of the run-down is controlled by the timing capacitor selected and by the setting of RV51. The run-down is 'caught' at 106 volts by V52B, preventing jitter which might otherwise occur when the valve 'bottoms'. At the end of the run-down, the Supertron regeneratively resumes its stable state; this causes a sudden fall in the screen potential, which is differentiated by C185 and R378, and triggers the time-base.

After triggering of the Supertron, the anode begins to run down.

The negative going waveform is coupled back to the grid through V54A and a timing capacitor. The timing capacitor is selected by Sw9/5.

This gives a 'Miller' type run-down. The speed of the run-down is controlled by:
(a)   the timing capacitor selected;
(b)   the setting of RV51.

The run-down is 'caught' at 106 volts by V52B. This prevents jitter which might otherwise occur when the valve 'bottoms'.

At the end of the run-down, the Supertron regeneratively resumes its stable state.

The resumption of this stable state causes a sudden fall in the screen potential, which is differentiated by C185 and R378. It also triggers the time-base.

But let me emphasize that this style should be used sparingly — for *parts* of texts only. Readers react strongly against the 'childishness' of whole books presented in this style. To present a mass of information that needs to be broken into many small steps, consider using integrated verbal and graphical presentation techniques, such as flow-charts and algorithms, to bring order, clarity, and variety to the presentation.

4.2 Specifications

Specifications must, above all, be unambiguous. This requires writers to be explicit about who is to do what, where, when, in what order, how, and with what materials. These are mainly matters of content: but style also needs care. An ill-chosen word can leave a gap through which much time, money and goodwill can leak away. The following extracts from specifications exemplify the most common weaknesses that are matters of style:

KILN INSTRUMENTATION AND CONTROL SPECIFICATIONS

1. Each pair of kilns will be provided with the following control equipment:
4. The instrumentation and control scheme shall be designed and
6. X indicators shall match the controllers and recorders, if possible.
12. must be capable of withstanding an overload of approximately 10X range
14. Pneumatic supplies are required for cylinder actuators.
29. Scan-fold chart design is preferred.
42. Prices should be shown ex factory.
52. It would be advisable for X Co [the contractors receiving the specifications] to have insurances in force to cover liability for

Item 1 introduces one of the commonest weaknesses in the writing of specifications — the use of *will*. Unfortunately, in casual day-to-day usage in British English, we use *will* in three ways:

1. as the third person form of the future tense of the verb *to be,* to make a simple statement of something that will happen in the future:

'The reactor will be built in two stages'
'The frame will be constructed in 3 in. tubular steel';

2. to express determination:

'We will build the reactor, in spite of local opposition'.
'I will construct the frame in 3 in. tubular steel, although you disapprove';

3. to express obligation:

'The contractor will ensure that the reactor is complete by'
'The frame will be constructed in 3 in. tubular steel'.

None of these uses is 'wrong'. In general, a word's 'meaning' is what most people use it to mean, and even though a word is frequently used in several different ways, provided the context makes clear the speaker's intention, there is rarely any ambiguity or misunderstanding. But the differences in meaning between uses 1 and 3 above are signalled mainly by intonation and context. In specifications, we cannot leave so much to chance. As far as possible, we must use words in a way that will leave no room for varieties of interpretation, and at points where we want our text to have mandatory force, we must choose words with particular care.

In British contexts, *will* used in third-person constructions is not regularly recognized as having mandatory force. It would be possible for a contractor to argue that he understood Item 1 to imply that someone else was to provide the kilns with the control equipment.

The word *shall* is more reliable. Although *shall*, too, is used in a variety of ways, similar to the uses of *will*, in British contexts *shall* is usually accepted as having mandatory force in third-person constructions. Though Item 1 might be challenged, the obligation intended by Item 4 would not normally be open to question.

However, writers of specifications would be wise to avoid all possibilities of confusion by avoiding both *will* and *shall* as means of expressing obligation and liability. It is possible to do so by using *must* or *is/are to be* instead.

Items 6 and 29 demonstrate two common weaknesses in specifications: the addition of the words *if possible* and the use of the expression *prefer(red)*. Writers of specifications should make up their minds whether or not features or qualities are essential. There can be no complaint if expressed preferences are not met. If certain features or qualities are essential, there should be no option left open; if they are not essential, there is no value in mentioning them in the main specification. Of course, it may be true that the features or qualities are *not* vital but are desirable. In those circumstances, it is preferable to state that fact outside the main specification or at least in notes set in a distinctive way separate from the obligatory items of the main specification.

Item 6 is typical in another way, too: in its careless use of the word *match*. Does this imply equivalence in size, shape, manufacturer, material, colour? Words like *similar, like, matching,* and *equivalent* are all dangerous when used in specifications.

Item 12 illustrates a trap that writers of specifications and of operating instructions often stumble into: inexact specification of quantities or values. What variation is permitted by the phrase: 'approximately 10 times a range'? Plus or minus 0·5? Plus or minus 1·0? Plus or minus 1·5? Almost always, a safer way of indicating quantities or values is to specify 'not more than', 'not less than' or 'between'.

Item 14 does not specify. The form *are required* is as weak as *will*, as it simply states or comments on a need. The item should read: 'Pneumatic supplies *must be/are to be* provided for the cylinder actuators'.

Finally, Items 42 and 52 both apparently leave options open again. If matters are optional, they should be distinguished clearly from obligations. Writers may argue that the use of *should/would* or *preferably* does distinguish options from obligations: I accept that argument in theory, but would contend that in practice it is wiser to separate the listing or options and obligations, either in separate documents or by distinctive layout within one document. By allowing themselves to use *should* and *would*, writers often carelessly express items as options when they really want them as obligations.

5 Style for Correspondence

In most correspondence, scientists are concerned with passing on administrative decisions or test results, with reporting new findings, with answering customers' enquiries, with following-up contacts that have been made by Sales Departments, or with discussing matters relating to materials and machines they possess or want. The aim is usually to transmit information exactly or to make a precise enquiry. There are, of course, occasions on which it is necessary to write a letter of apology, explanation, or conciliation; it may be necessary to produce letters of complaint, letters presenting bad news, letters reminding friends and enemies of obligations; letters passing on congratulations and letters of commiseration. There is also widespread activity in what might be described as sales writing — the use of letters to circulate information about products and processes. But these types of correspondence are not the normal day-to-day concern of scientists and engineers, so I propose to confine my discussion of letter-writing to the comparatively straightforward tasks of giving or seeking scientific information.

The aim of scientific correspondence is normally to transmit information directly and specifically: accordingly, writing should be direct and specific, and much of the advice given earlier in this book applies again here. But there is an additional factor to bear in mind. That is, that letters help to create 'the company image' of organizations. Everything we write not only passes information but also creates an impression of ourselves in the readers' minds. High-flown style creates an impression of pomposity; roundabout phrasing creates an impression of tortuous thinking; impersonal, passive writing creates an impression of inert facelessness. A business letter is written on behalf of a company or organization and, inevitably, the character that suffuses the writing is partly transferred to the organization.

Do you want to give the impression of a cold, ponderous

juggernaut, grinding impersonally and unswervingly along immutable tracks? If so, write something like this:

Receipt of your letter reference AB/CD/1234 dated 14th inst. is acknowledged with thanks.

It is regretted that this Company is unable to comply with your request for X-ray machinery having dimensions within the limits specified in your enquiry. The standard apparatus for diagnostic work produced by this Company is . . . . It is possible, however, that a special machine could be produced, capable of fulfilling your requirements. Any order for such a special machine would necessarily involve a substantial increase in costs.

This is 'correspondence English' in its most deadening form. Our aim should surely be to create the impression of an efficient, friendly organization of human beings, in which there are technical specialists interested in finding solutions to the customer's problems. This calls for a clear, friendly, unaffected approach:

Our usual X-ray machines would be too big to fit into the space you have available; but we could certainly make a special machine

The features of style that I have stressed already in this book apply as much in letter-writing as in any other writing. Directness and activeness should be the keynote:

Write: I enclose a sample of XY 123 for you to evaluate
not: A sample of XY 123 is enclosed for your evaluation

Write: We are developing this method of spraying
not: This method of spraying is actively receiving our attention with a view to further development.

Write: We suggest you try to overcome this difficulty by
not: It is recommended that an attempt should be made to obviate this difficulty by

In a serious conversation with a customer or contact on the telephone, you would not say: 'It would be appreciated if you would favour us with your comments': the comfortable rhythms and vocabulary of direct speech would produce something like 'Please let us know what you think'. These words are easier for the receiver to assimilate; there is nothing inaccurate, slangy, or discourteous in them; they are therefore preferable in writing as well as in speech.

I am not for one moment advocating the use of loose, slangy, chatty diction. My attack is as much against the jarring effect of sudden lapses into racy 'mateyness', which offends by its presumption and imprecision, as against the blurring effect of over-formal phrasing or unnecessary jargon.

Try to think that your client or customer is sitting right in front of you; then write as far as possible in the words you would use in a full, direct and precise account of your message. In this way, you should avoid erecting an unnecessary barrier of 'correspondence words' between your customer and your message.

Correspondence words are words which somehow seem to take over from perfectly good everyday words as soon as people begin to write letters:

Use	Don't use
tell	acquaint/advise/inform
begin/start	initiate/commence
send	despatch
buy	purchase
soon	at an early date
because	due to the fact that
we can	we are in a position to undertake
about	concerning/regarding/with respect to/in connection with/in relation to/with regard to

These words and phrases are undesirable *not* because they cannot be understood, but because they build up an unnatural, stiff, unnecessarily formal atmosphere about letters.

Why do writers use this style? It seems that they believe that letter-writing calls for greater dignity or formality than discussion or conversation. This is almost true, but not quite. In writing, we certainly need to be *more explicit*. We have to put into words some elements of our message which in speech we could rely on passing by intonation, gesture or other signal. Also, we have to be sure that the form of words we use in writing will be entirely comprehensible to all our readers without ambiguity, and without further explanation. In contrast, in speech we can try out one form of statement, and if we see that it has not been comprehended, we can rephrase it immediately, expanding it and reorganizing it until our listeners understand.

Writing calls, then, for greater precision and greater fullness of detail: but it does *not* call for a more roundabout, more pompous, more long-winded style.

To keep letters as natural and direct as possible, use the first person, singular or plural, whenever it seems comfortable to do so. This does not lay you open to personal responsibility for the information you give; use of the first person will not make you more or less responsible. You write as the company's agent anyway; you use notepaper headed with the company's name and give your office in the company after your signature. If you sign a letter which makes a recommendation, you will not decrease your personal involvement by using the more awkward, evasive phrase 'It is recommended that'. Other avoiding phrases such as 'the undersigned' or 'this office' or 'the present writer' are just as awkward:

> The writer is aware of the use you have made of A, B and C for various lubricating purposes over the past years, but we would advise you that X will fulfil all the functions hitherto performed by...

> Therefore, you are requested to communicate initially with the undersigned at this office.

Notice how the sentence structure of the first example is contorted to accommodate 'the writer's' unnecessary attempt at self-effacement. To me, the unnatural construction 'the writer is aware of the use you have made' is far more obtrusive than the simple 'I know you have used' would have been. Similarly, in the second example, the artificial expression 'communicate with the undersigned' jolts me into noticing the style used in the writing, when I ought to be concentrating simply on the message.

Probably the most natural way of writing is to use the first person plural, emphasizing that you are a member of a company of lively people. You can then vary your constructions and use the first person singular or the impersonal to gain variety or particular effect:

> We have three experimental constructions set up to test these procedures. If it would help you to come to discuss these with us, I should be pleased to see you any afternoon next week.

One further point we must bear in mind is ensuring that we give the right impression of courtesy. Naturally, in writing to customers we must pay particular attention to being courteous. Avoid being too abrupt, commanding or supercilious:

You did not send us a large enough sample to test fully. Please ensure that samples sent for test in future are a minimum of

or sounding tactlessly critical:

It was unreasonable to hope that such a pressure could be achieved in your existing steam-producing plant

You must have misunderstood our instructions about

But this does not mean that you should go to the other extreme and become unnaturally deferential or obsequious:

We would respectfully beg to inform you

accept our humble apologies for the unfortunate incident and trust that this will not disturb the confidence hitherto reposed in us should we be given a further opportunity to

Avoid, too, the inflating effect of artificial gush and dramatic flourish:

It is with great pleasure that we have to inform you
We would venture to suggest to you

Excessive courtesy rings false and hollow sentiment is much worse than none at all.

Some of the most awkward writing in correspondence comes in the opening paragraphs:

Adverting to your esteemed order of 24th inst. concerning nuts and bolts

Re your letter of yesterday's date, reference AB/CD/1234,

Further to your telephone communication of yesterday

Thank you for your letter of 24th January 1985, reference AB/CD/12345, regarding the position in respect of your order for aluminium alloy window frames

We enclose herewith our catalogue re our telephone conversation of yesterday's date entitled 'Polyacrylate rubber'

An excuse that is often offered for such clumsy opening sentences is that it is essential to mention the date, the reference number, and the subject of the customer's letter as early as possible in the first sentence. This is not necessary in a well organized letter. The customer's reference number should be put at the top of your notepaper, against a printed sub-heading 'Your reference'. This reference will often include the date and an indication of the subject of the letter; but if you are still anxious to make the subject plain, perhaps because you write many letters to the same customer on various subjects, put a heading immediately after 'Dear . . . . .'.

Dear Mr Brown,

Aluminium Alloy Window Frames

This leaves only the date of the customer's letter to be mentioned in the opening sentence and this can usually be done unobtrusively:

Half the nuts and bolts you ordered on the 24th January are now ready and

There are two possible solutions to the problem you raised in your letter of 24th January

After our telephone conversation yesterday, I checked

Above all, don't set your letter off to a bad start by opening with a grammatical error:

With reference to your letter of 4th January. We are sorry

In reply to your letter of 7th December. There is still time

These are incomplete sentences. If you open with 'In reply to', you must go on to state what you are doing or have done. Begin in a crisp, businesslike way with your first point. Avoid the stereotyped, cumbersome opening:

Thank you for your letter of 28th June 1984, reference AM/Trg/14, and your enquiry regarding the possibility of our presenting Effective Communication courses in connection with your company at the Grand Hotel, Seaside. I am writing to inform you that negotiations with that hotel have now been brought to a successful conclusion and we are able

Begin with the first point you wish to make, particularly reserving the opening 'Thank you' for occasions on which you really wish to thank people for a particular service or favour:

Negotiations with the Grand Hotel, Seaside, have now been completed and the only matter still to be decided is

Thank you for sending me an extra copy of the agreement on

Would it be possible to change the date...? One of our staff is seriously ill, and

Finally, you have to end your letter. Don't feel that you must finish with a flourish. If you use one of the customary formulae, your flourish will be only too plainly an empty gesture:

We trust that this information will be of use to you, and look forward to being of future service

If you would like further information, please do not hesitate to contact us

Assuring you of our best attention at all times

We await the pleasure of your further instructions

Yours in anticipation of an early reply

We look forward to hearing from you at your convenience

The point about these sentences is that they are now so over-worked that they are merely empty noises. If you really want a particularly prompt answer, ask for it directly and say why. Otherwise, avoid the coercing tone of the last two examples (and the probable snigger at the double meaning of the last). It is much better, when you have finished your message, simply to stop. Most people will not feel that this is unduly abrupt: they are more likely to get a sense of clear thought and crisp control.

This discussion has been concerned with letters written in response to a previous contact — to answer an enquiry, to give information following a test, or to comment on a suggestion. These will probably be the main part of your day-to-day correspondence. But sometimes you will be a customer or enquirer yourself. The need for care and exactness is just as great. Make plain what you want to know. A vague question invites a vague answer. It would be difficult to give a direct and helpful answer if you asked a question such as:

I am interested in switches for a new thermostatically controlled electric radiator. Please send me details of any appropriate products in your range.

6 Writing for Users of English as a Foreign Language

6.1 English for International Communication

The theme of this book so far has been the need for scientists to write accurately, directly, and sensitively — with *sensitively* meaning 'with due attention not only to the accuracy but also to the manageability, readability and acceptability of what they write'. I have stressed that this entails using language flexibly, selecting from the widest possible *variety* of vocabulary and structures just the right language to give the desired meaning, balance and tone. In particular, I have condemned the distorting effect produced by efforts to straitjacket scientific ideas entirely within impersonal, passive, past-tense constructions.

But there are circumstances in which it *can* be beneficial to restrict the forms of English used in writing. In documents that are to be read mainly or largely by people for whom English is a foreign language, careful control of the range of language can help the readers considerably.

Native speakers of English must recognize that delicate distinctions of meaning or subtle shades of tone may not be recognized by people for whom English is a foreign language. Speakers of English as a foreign language may not understand that the phrase 'a *novel* technique' is intended to convey more than that the technique is new: they may miss the intended implication that it is 'ingenious, rather clever and interesting' as well as new. Readers using English as a foreign language may find it difficult to recognize that the statement 'production is to all intents and purposes static, tending if anything to decline' is intended to mean 'production is falling slightly'. They can be forgiven for not knowing that 'it could be said that' usually is a self-protective way of saying 'I think': the use of *could* certainly does not imply a particular condition. And they are almost certain to have difficulty in grasping the meaning of the sentences below — most English-native readers do, too:

The contribution of the series resistance of the electrolyte towards dissipation factor decreases with decrease in capacitance (for the same case size), therefore there are lower losses in capacitors with the higher rated voltages.

Since the degree of concavity is an inverse function of the degree of separation, and capillarity influences its height without altering its basic character, it follows that 'capillary manifestations in concave variants' is the correct description and prediction (generalization) of the character and behaviour of a sinal fluid level, and that these fluid levels are co-variant with the characteristics of space at the level of sinal pooling in all erect positions of the skull.

Authors may think they are showing a praiseworthy command of English if they make extensive use of statements intended to imply more than they say on the surface, or if they make extensive use of verbs such as *should, would, could, may* and *might* to impute varying degrees of certainty, or if they make extensive use of complicated sentence patterns. But extensive use of these features of language is not likely to impress readers who are using English as a foreign language. For such readers, these features are more likely to cause difficulty and confusion. For effective *international* communication, especially for communication in commerce, industry or research, it is wise to restrict the range and the complexity of the English used.

Indeed, it is possible to restrict the English used in written documents so carefully that the documents can be 'read' by people who do not truly 'understand' English at all. These people do not 'read' English: they simply recognize patterns of familiar symbols, or marks and lines on paper. Yet, they can be taught to work efficiently from documents written in this restricted range of symbols, in this 'controlled' English. There are obvious commercial advantages in being able to distribute internationally just *one* controlled-English version of a piece of information, instead of having to publish many versions in different languages.

Of course, it is necessary to use a complete range of English to express complex or abstract ideas. Controlled English *can* cope with commercial and technical information like installation instructions, maintenance and repair instructions, operating procedures, programme writing, and various types of descriptive

writing.   Controlled English can *not* cope with theoretical discussions, arguments about data, or with very abstract analyses.
This means that controlled English cannot cope with much of
the complex, new scientific and technical information that must
be distributed internationally: that information calls for use of
much wider language resources.   Nevertheless, even if we have
to present complex information, we should keep the principles
of control clearly in mind as we prepare to communicate with
international audiences.

6.2 Writing for Expert Readers

In international communication in commerce, industry and
research, there are three broad groups within the audience.   For
convenience, I shall call them the expert group, the student
group, and the no-English group.   (I should stress here that I am
confining my discussion to international communication in
commercial or professional activities.   I am not discussing
'English for tourism' or the development of communication
skills for general social and cultural activity, which would
introduce other groups with other needs and motives.)

The *expert* group is perhaps the largest of the three.   It
consists of:

1. people who want to read international journals and books
   about science, technology, commerce or legal matters;
2. people who are working with internationally available complex equipment;
3. people who are involved in international commercial activities.

These people read books or listen to lectures in English because
they already have some knowledge or expertise in the subject-
matter of the books, the operation of the equipment, or the
activities of the international organizations with which they are
connected.   They have learned their fundamental conceptual
knowledge in their own language or in English.   For many, the
specialist vocabulary they have learned is the same in English
and in their native languages.   Those who are not so fortunate,
have to learn the equivalent specialist words in English and their
own languages.   This is hard work, but it is not usually confus-
ing.   There are usually reliable direct translations between the
two languages.   For expert readers like these people, it is *not*

usually the specialized English vocabulary that causes trouble. Either they know the words, or they can look for the meaning in a dictionary. For readers at this level, writers can reasonably assume that the audience can understand the *established core* of specialist vocabulary.

I am well aware that it is not easy to describe this established core exactly. I would certainly agree that writers should explain slightly too much rather than slightly too little. My point is that for this expert audience, the most frequent causes of difficulty are not single specialist words but complexities of sentence structure, verb forms and apparently *common* English words and phrases that English native speakers use almost without thought.

Consider, for example, the following paragraph:

Sentence 1	In such a case, the program would be run once and provide one set of results.
Sentence 2	In fact, it is the simplest form of program possible except for one which does not have a repetitive loop.
Sentence 3	To make the program more useful still, we should have to put in an input section in which we could read in different sets of values of A and B.
Sentence 4	This would necessitate the use of another loop.
Sentence 5	We might then decide to add the facility of changing the range of X and the X increment.
Sentence 6	This would introduce another input step and another loop.
Sentence 7	The possibilities are endless.

People who read this text must know a great deal about computing already, otherwise they would not be reading it. For them, specialist words like *program, repetitive loop, input,* and *increment* are not likely to cause difficulty. Much more difficulty is likely to come from the mixture of positives and negatives in the second sentence 'simplest ... except ... not'. More difficulty is likely to come in the third sentence from the phrase 'more useful still' and from the shades of meaning of *should* and *could*. More difficulty is likely to come from the optional use of *might* in sentence 5. And some will find it difficult to comprehend the final idiom 'the possibilities are endless'.

Again, in the following extract, the main source of difficulty is not likely to be the specialist terminology:

At full loadings the output voltage and current are the combination of two components in quadrature, one from the secondary winding of the transformer T.2 and the other from the secondary winding of the transformer T.1. With this circuit, if for a given value of load resistance R, the reactance of the transformer T.2 is minus R and that of T.1 plus R, then the load current of T.1 is the reactance current of T.2 and the load current of T.2 the reactance current of T.1, thus the former leads the supply voltage by 45° and the latter lags by 45° and the two load components are thereby 90° displaced. The output has a two phase character, whilst the input current is in phase with the supply voltage.

In this example, the greatest source of difficulty, I suggest, is the unnecessary complexity of the sentence structure.

The following examples again show difficulties arising not from the specialist vocabulary but from thoughtless writing. Here, the examples show writers not recognizing the difficulties caused for overseas readers by informal phrases like 'will remain tied up':

if all aircraft are stored with tanks two-thirds full, vast quantities of fuel *will remain tied up* for long periods

the X Doppler system *is thrown in* mainly to

if the system *does not come up with* your requirements

the suggested delivery dates *will be tight*

but since the X system is *quite* new
(meaning *entirely*)

Writers in English must even take care about the words and expressions they use in communication between England and the USA. The Englishman who is visiting a laboratory in the USA and is asked 'What time do you have?' will find his reply 'Not much, I'm afraid' greeted with some confusion. His answer should be based on a quick reading of his watch — 'Ten minutes past three' or 'Ten after three'.

It is important to remember that there is not just one English language: there are (at least) British English, US English, Canadian English, Australian English, and South African English.

These languages are marked principally by similarities; but there are many differences, too. If we want high reliability in the transmission of information and if we wish to avoid causing offence by creating inept tone, we must ensure either that all our documents are written in the language of the target audience or that the target audience is fully alerted to the points of difference in the code that is being used.

It is interesting to recognize that it is not the big and obvious differences that cause difficulties in transatlantic communication. For example, pronunciation and intonation are obviously very different in British English and US English, but after a tuning-in period, we do not often find that pronunciation and intonation differences cause complete breakdown in communication.

There are some differences in labels that must be learned: for example, US *faucet*/British *tap*, US *ground*/British *earth*, US *thumb tacks*/British *drawing pins*, US *antenna*/British *aerial*, US *hood*/British *bonnet*, US *in back of*/British *behind*, US *initiate your main lamps*/British *switch on your headlights*: but we soon learn these, because if we do not, communication stops.

Much more dangerous are apparently common idioms or words with which we assume everyone on the receiving end is familiar: communication goes on, with participants unaware that there are differences of interpretation between transmitter and receiver:

US   use *alternate* equipments
     (British : alternative)

US   drive on the *pavement*
     (British : drive on the road)

US   the *regular* operator
     (British : the usual/ordinary operator)

But he was not *about to go forward* with his plan
(US : he had no intention/was not likely to go forward)
(British : he was not on the point of going forward)

Have them inspect the sample first
(US : first, show them the sample)
(British : compel them to look at the sample first)

Notify the Supervisor when *you are through*
(US : when you have finished)
(British : when you have penetrated)

There can be difficulties over the use of abbreviations, too: for example, over the American use of # to mean number, over the European use of the comma as a decimal point, and over the American way of indicating dates (3/12/84 = 12th March 1984).

So, for expert readers overseas, even in the USA, writers should constantly keep in mind the nature and range of English likely to be at the disposal of the audience. It will almost certainly be desirable to limit the language forms used and to make explicit many nuances of meaning and tone that could be expressed to native English readers by subtleties of structure or vocabulary.

6.3 Writing for Students

In the student group within the international audience, I include all people who are studying English itself or who are using English as the medium of study of commercial, scientific or technical subjects. These people have not learned the fundamentals of their subjects in their own languages. They have to build up in English the whole conceptual framework underlying each special term. When we plan to write or speak to such readers, we must make great efforts to give clear definitions and explanations.

This does not require different skills from those we need for writing for expert readers. The task differs more in degree of difficulty than in kind. The main difficulty is still to control carefully the complexity of the structures created. We must make even greater efforts to control the language used and the rate at which information is unloaded. The greatest difficulties for student readers are again caused by difficult phrases, complex sentences and shades of meaning expressed by *would, should, may* and *might*.

6.4 Writing for Readers Who Do Not Understand English

When we come to present information to the third category in our audience, to the no-English group, the group who do not know how to speak or write *any* English, we have to exercise the greatest restriction and control. Yet, and this may seem strange, provided we know that these readers have been trained to recognize the words (symbols) we use, we can be more confident about communicating clearly with these readers than we can about communicating clearly with either of the other groups. When we communicate with experts or students we can never be absolutely sure that we have judged accurately their ability to understand us. In contrast, if we know our readers have been taught to recognize the symbols we use, we can be confident of clear communication.

Several 'controlled' versions of English have been developed by large international companies like the Caterpillar Tractor Company, Digital Equipment Corporation, L. M. Ericsson, IBM, Eastman Kodak, and Rank Xerox. All are based on an original idea worked out by the Caterpillar Tractor Company of Peoria, Illinois.

The Caterpillar Tractor Company distributes tractors and heavy earth-moving equipment throughout the world. It supports its machinery with maintenance and repair documents, and because this has to be done in many countries, the company used to provide the documents in many languages. But duplication of documents in many languages is expensive, so the company looked for alternative ways of presenting its information.

Caterpillar research workers decided that it should be possible to use a single, internationally understood set of symbols to convey much of the information they had to transmit. And having decided this, they explored the possibility of using English words as this set of symbols. They produced a list of 784 words as a central core of symbols, plus a list (with many illustrations) of names of parts of Caterpillar equipment.

The research workers found that they could express all their service and maintenance information using this vocabulary alone in a carefully controlled range of simple structures. They found, too, that in 30-60 hours they could train operators *who previously knew no English* to recognize the meaning of the documents written in this way. Though the operators did not 'understand' English, they could work efficiently on the basis of the information drawn from the controlled-English documents.

The Caterpillar Tractor Company named its restricted version of English 'Caterpillar Fundamental English'. It is marketed now as ILSAM — International Language for Service and Maintenance — by M. and E. White Consultants (world-wide agents) and as BASIC 800 by Smart Communications Inc. (agents within the USA only).

The principles for producing a restricted or controlled version of English such as ILSAM or BASIC 800 are easy to grasp. The variety of words used must be strictly limited and each word must have one meaning only. For example: *right* is the opposite of *left*, *correct* is the opposite of *wrong*. *Drop* is a noun meaning 'a quantity of fluid that falls in one spherical mass'; it is not used as a verb meaning 'to fall' or 'to release'; and it is not used as a noun as in 'a drop in pressure' — that is a *decrease*. The word *over* is restricted to the single meaning 'above', as in 'placed over the valve'; it is not used to mean 'more than' as in 'produces over 10 watts', or 'during' as in 'over the three days', or 'finished' as in 'that the emergency was over'.

Synonyms are avoided: from several words that have approximately equal meaning, one only is chosen for use. For example: *below, under, beneath* and *underneath* all have similar meaning. ILSAM uses only *under*. Also, wherever possible, the word with the widest international recognition is chosen for use: *assistance* is used, not *help*.

The number of verbs is kept to a minimum. This is achieved by use of verb-noun combinations as much as possible: *make an alignment* is preferred to *align*. This has the virtue of reducing the number of verbs to be learned and of reducing the number of irregular verb forms to be used.

Statements are made as short and positive as possible.

NOT: The control unit, duplicated for safety, has a low resistance.

BUT: The control unit has a low resistance. There are two units for safety.

Repetition replaces reference back and explanations are carefully sequenced in steps. Tense, voice and mood of verbs are carefully restricted. Wherever possible, sequences of statements plus words like *before, after, last, first, then* are used instead of past and future verb forms.

Writing is restricted to four types of sentence structure:

1. statements, descriptions or explanations;
2. instructions or commands;
3. combinations of conditions with either descriptions or instructions;
4. questions.

Statements can have qualifying words added, but not too many. If necessary, adjectives can precede both subject and object. The first sentence below has no adjectives: the remaining sentences add more:

The washers prevent leaks.
The network uses reed switches.
The control system has two complete stages.
Twelve GV-1 groups are normal in group selectors.

Heavily qualified statements are avoided.

In instructions or commands, each sentence contains just one instruction:

NOT: After stopping the program, load the data into the buffer store.
BUT: Stop the program. Load the data into the buffer store.

NOT: Using program 6, send X to Y.
BUT: Use program 6. Send X to Y.

Clauses giving explanations, conditions and indications of time can be combined with descriptive statements or instructions:

Reason clause: Because you must make the loop first, the linkage is important.

Time clause: When the work is complete, put the test line into the cable duct.

Condition and instruction: If the error is larger, increase input at X.

Condition and statement: If the interval is less than three seconds, the sequence is wrong.

But these combinations must remain manageable: one qualifying clause per statement is a good rule.

Questions are constructed as simply and directly as possible:

Can the door open freely?

Of course, non-English-speaking operators cannot just pick up and read documents in ILSAM or any other controlled version of English, however carefully the documents are written. A training programme is necessary. In the training programme, a bilingual instructor helps the operators to recognize and understand the significance of the limited vocabulary and the limited range of patterns. He gives definitions and explanations in the native language. But at no point is the learner required to speak or write English. Caterpillar have found that a course lasting 30-60 hours is normally enough to enable 'readers' to work competently from ILSAM documents.

An important feature of this type of controlled English is that it is not distorted or artificial. The vocabulary and structures it uses are selected from those used daily by native English speakers. Documents in this language are therefore entirely acceptable to native English readers. Indeed, when the Caterpillar Company published the first service literature in its restricted language, the difference was not detected by native English readers!

Here are some examples of texts re-written in 'Ericsson English' — a controlled version of English produced by UWIST for the L. M. Ericsson Telephone Company, Stockholm[1,2,3]. The verbs are in italics. The original texts use relatively complicated verb forms such as 'have been positioned' or 'being untwisted', and the auxiliary verbs 'may' and 'should'. Passive constructions and auxiliary verb forms are major sources of difficulty for readers who are not proficient in English and they are not necessary in simple instructions and explanations. Note, too, how it is possible to have different re-written versions: use of controlled English does not mean that writers have no choice in the language they use.

Original text

Stripping of cables for normal magazines

When the cable *has been positioned* on the cable shelf for normal magazines, *use* measuring rib 1007478 to mark where it *is to be* stripped. Stripping *is to take* place immediately before connection work as the twisted groups *run* the risk of *being untwisted* if the cable *is stripped* in advance.

One re-written version

Stripping of cables for normal magazines

Put the cable into the correct position on the cable shelf. *Use* measuring rib 1007478 *to make* a mark on the cable. The mark *shows* where to *start* stripping. *Strip* cables a short time before connecting, *to prevent* untwisting of cable elements by accident.

Another re-written version

Stripping of cables for normal magazines

Make sure that the cable *is* in the correct position on the cable shelf for normal magazines. *Use* the measuring rib 1007478 *to make* a mark on the cable, so that the mark *shows* where *to strip* the cable. *Do not strip* the cable until you *are* ready to start connecting. If you *strip* the cable before you *are* ready, you *can untwist* the twisted groups by accident.

Original text

In order *to be able to keep* the time schedule, it *is* important that the preparations *are* well *made* and carefully *thought* over (items 3.2 and 3.3), otherwise a waste of time *will* easily *ensue*. As *is* evident from the suggestion for time schedule below, several different tests *may be* in progress simultaneously. Just how long a time each test *will take is* dependent on local circumstances, but no test, with the exception of certain APZ tests, *should take* more than 8 hours unless interferences *occur*.

One re-written version

We *make* careful preparations according to Sections 3.2 and 3.3, so that we *use* the period of the operation test in the best possible way. Figure X *shows* a plan of the length and sequence of the tests. The length of each test *depends* on local conditions. Several tests *can operate* at the same time. If we *do not find* faults, most tests *operate* for less than 8 hours. Some APZ tests *operate* for more than 8 hours.

Another re-written version

Important! *Be* careful when you *make* the preparations for the operation test (Sections 3.2 and 3.3). If you *are not* careful, you *will not use* all the available time successfully. Figure X *shows* that several different tests *can happen* together. The length of each test *depends* on local conditions. If faults *do not happen,* the maximum time necessary for a test *is* normally 8 hours. Some APZ tests *need* more time.

Original text

When the two bi-metal springs *have* room temperature a gap *should be provided*, however, maximum 0·1 mm, between the lower side of the link and contact spring 3. The distance *is measured* using a feeler gauge. Adjusting *is executed* by bending the adjusting tongues on the stiffened spring 4 using spring bender LSH 2602 or 2603. After adjusting a check *should be made* of the simultaneousness of the twin contacts. No bending of the bi-metal springs *must be made.*

One re-written version

Let the temperature of the two-metal springs *become* the same as room temperature. The springs *will* then *open.* The maximum gap acceptable between the lower side of the link and contact spring 3 *is* 0·1 mm. *Measure* the gap with a feeler gauge. If you *need* to adjust the springs, *use* spring bender LSH 2602 or LSH 2603. *Bend* the adjusting tongues on spring 4. *Do not bend* the springs. Then *check* that the two contacts *operate* together.

Another re-written version

Let the temperature of the two two-metal springs *become* equal to the temperature of the environment. *Use* the feeler gauge *to measure* the gap between the lower side of the link and contact spring 3. The maximum acceptable gap *is* 0·1 mm. If the gap *is* too large, *use* a spring bender LSH 2602 or LSH 2603. *Bend* the adjusting tongues on the rigid spring 4 until the gap *is* small enough. *Do not bend* the two-metal springs. *Check* that the contacts still *operate* correctly.

Even if a company feels that it cannot use a single controlled-English version of its document(s) in *all* countries of the world

— that is, that some translations are still wanted for diplomatic or technical reasons — the creation of controlled-English text(s) still has value. Since a controlled-English text has been written very clearly, with each word confined to just one meaning, it is ideal source material for translation. The translator can rely on the definitions of words and finds the simple sentence structure relatively easy to convert into comparably simple structures in the target language. Indeed, a reasonably accurate 'base' translation can be produced easily by computer, leaving human translators to add the expert touches — to add balance and polish by making the adjustments of word order and style required by the target language.

Principles similar to those used in creating ILSAM have been used to create controlled versions of English for international oral communication. For example, there are standard patterns of English for international aviation, and a Standard Marine Navigation Vocabulary has been produced by IMCO (Intergovernmental Maritime Consultative Organisation)[4].

The aim of the IMCO Vocabulary is to encourage all navigators to communicate effectively with one another by making them use an agreed range of words and phrases with agreed meanings. The Vocabulary suggests words and phrases for use in a variety of navigational activities. The words and phrases are arranged in groups under 21 activities such as anchoring, fairway navigation, pilotage, radar transmission and fishing. There is a glossary explaining difficult terms, and an introduction describing how the Vocabulary should be used.

The design and use of controlled versions of English primarily (though not exclusively) for oral communication raise problems related to speaking and listening that do not concern us here. However, readers may be interested to explore the potential of controlled versions of English for international communication of commercial, scientific, and technical information in both speaking and writing. There is considerable scope for extension of the principles already used in technical documentation and in aviation and maritime activity.

References

1. Kirkman J., Snow C. and Watson I., *Controlled English in International Documentation*, Proceedings of the 39th Congress of the International Federation for Documentation, Edinburgh, 25-28 September 1978.
2. Ditto.
3. Kirkman J., Snow C. and Watson I., *Controlled English as an Alternative to Multiple Translations*, IEEE Transactions on Professional Communication, Vol. PC-21, No.4, 1978, pp.159-161.
4. Department of Trade, Merchant Shipping Notice No. M702, Standard Marine Navigational Vocabulary, London, 1974.

7 On Avoiding Ambiguity

I am often asked for advice on how to avoid ambiguity in scientific writing. One answer I can offer is the use of carefully controlled language, limited in vocabulary and structures, as described in Chapter 6. But such a form of 'controlled' English will not meet all our needs in writing scientific papers. To communicate the shades of meaning and nuances of tone in subtle and complex description and argument, we need to use the full range of the resources of English. In such circumstances, I have to give a different answer to the question about avoiding ambiguity. The answer is: we can't.

As readers or listeners we can, if we wish, wilfully misconstrue almost any statement, especially one which contains general rather than special vocabulary. In reading an article about new designs for trains and stations, we may smile momentarily at the double meaning of the statement 'the buffer will have an inflated head'. But it is really *our own* fault if we are distracted from the intended information about a mechanical device by an image of a human being with a peculiar physical shape.

Normally, we do not decode messages by looking for the *least* likely meanings that can be attached to the signals we receive. We do not receive and interpret the signals in isolation: we take into account the whole context in which they are being used. We consider the general linguistic surroundings of the particular signals; we consider the general theme that is being discussed or the total object that is being described; we consider the writer's personality, education, occupation and experience, and the atmosphere pervading the communication situation as a whole. All these things are taken into account as we consider possible interpretations. This means that sentences including the following groups of words, which some people would condemn as ambiguous, would normally give no trouble:

> acquires a pitch that is in proportion to the distance the bat is away from the reflecting object
>
> these stages consist basically of staircase generators

The meanings we draw from *pitch* and *bat* vary according to whether they are included in a discussion of ultrasonic guidance systems (as here) or a sports report; and indeed their sporting meanings vary according to the game being discussed and whether the context is British or American. No reader is likely to be seriously confused. Likewise, no reader of the electronics journal from which the second extract was taken would deduce that the stages consist of machines busily producing staircases!

So in this chapter I want to stress the impossibility of avoiding ambiguity *entirely* in the use of words. But let me make clear what type of ambiguity I am referring to when I make that statement: I am referring to the ambiguity that arises when we can make any of several possible responses to correctly formed and correctly presented code signals. My purpose here is not to excuse the types of ambiguity that arise from 'common errors' in sentence construction, syntax and punctuation — that is, from ill-formed and ineptly presented code signals. For example, I am not excusing ambiguity which arises from misused participial constructions.

> The problem of contamination by bacteria and yeast was a major difficulty and after running for about two weeks the mammalian cells were found to be non-viable.

I am not condoning ambiguity which arises from careless use of pronouns:

> Samples were sent to Technical Service Department for evaluation and stored for shelf stability. They have reported good adhesion and corrosion resistance and suitable shelf stability.

And I am not accepting ambiguity which arises from slovenly punctuation:

> counts of the bacterial load in the air of piggeries. He observed that bacteria carrying dust particles decreased in concentration...

These are not difficulties of semantics: they are weaknesses in thinking or defects of self-discipline in the writers. All good texts on efficient writing give copious advice on how to overcome them.

Nor, when I acknowledge the unavoidability of ambiguity, am I suggesting that we must condone woolly communication, which is unclear because a writer is tautologous, irrelevant or deliberately evasive, or because he fails to be specific in his instructions:

should take steps to overcome
(what steps?)

My concern is to acknowledge the extent to which the communication process relies on the audience's ability to *disambiguate* what the writer is trying to say. It is unreasonable to expect complete freedom from ambiguity in the words we use in speech and writing. Constantly, we have to leave the receivers to decide which of a number of possible meanings are to be attached to the words according to their contexts. And even, on some occasions, when it would be possible to eliminate all chance of ambiguity, it is not desirable to do so, because it would involve a circumlocution that would unnecessarily lengthen (and even complicate) the receiver's task of interpretation.

For example, in scientific documents it is usually quite acceptable to use terms such as *pitch, bat, gate, channel,* a *homing* device, or an *elastic* beam without elaborate explanation to ensure that misinterpretation is impossible. The essential skill in effective communication is judging what will be the familiarity of the receiver with the subject-matter and its customary vocabulary. The judgement that has to be made is not 'could this word be wilfully misinterpreted?' but 'is this word *likely* to be misinterpreted?'

Let me stress the importance of making this judgement. The language we use must be carefully and continuously adjusted to the capacities of the audience. My point is that *often* the audience is well able to select from possible meanings of the words used; the audience expects to have to select as a normal part of the information-receiving process, and does so without conscious effort. It is therefore unreasonable to give the impression that the use of words that are open to more than one interpretation is *always* unacceptable.

Unfortunately, some text books on efficient writing give this impression and it has been a frequent source of unease to students I have taught. For example:

'Good diction means the absence of ambiguity, obscurity and misunderstanding'[1].

'To produce a sentence capable of one meaning only every time you write (or speak) a sentence must be your aim as a communicator'[2].

'Ambiguity is a deadly sin in writing or public speaking'[3].

'Language must be entirely clear to its readers. All distortions and ambiguities are abhorrent'[4].

'The scientific writer has an obligation to use words as accurately as he does numbers and symbols'[5].

I suspect that none of these text-book writers really believes that it is possible to avoid ambiguity completely; much that they write elsewhere in their books acknowledges the difficulty of being unavoidably explicit; but their Olympian tone makes the learner-writer uneasy. Such advice is, I am sure, responsible for many of the ponderous linguistic contortions that inexpert writers indulge in as they attempt to be impregnably exact, complete, and (ironically) clear.

We all know from experience how difficult it is to be sure that our readers will distinguish between *discrete* and *discreet,* between *sine* and *sign*; we know that it is difficult to be sure that all our readers understand what is meant by a *dead* valve, or by *chatter* in machines; we know that the scientifically acceptable word *aperture* is no easier to define than the common-or-garden non-scientific word *hole*. We know, too, that to use the string of words 'four plus four times three plus two squared' cannot be as clear as to use algebraic notation of numbers and symbols. So we must despair of being able to keep these text-book commandments.

The advice we give should surely be less dogmatic. We should acknowledge that it is difficult to express meaning exactly; we should point out that there are limits to the writer's responsibility for removing ambiguity; and we should suggest that the best way of minimizing ambiguity is to focus attention firmly on the needs and capacities of our readers.

Our difficulties in exact communication of our ideas arise partly because words have both connotative and denotative meanings. Louis Salomon defines these terms in the following way[6]:

> What we shall call the denotation of a word is the sum total of its referents: for example, *chair* denotes every single chair that has ever existed, or ever will exist, in the world of sensory experience. The connotation of a word is subdivided into two parts: the defining qualities of the category or class it names, and the emotive or affective responses it arouses in the minds of its users. In popular parlance 'denotation' is often thought of as the real or proper meaning of a word, and 'connotation' as the mere accretion of illogical, even capricious feelings that help to mask or colour this meaning — hence, something that a well organised linguistic society would probably legislate out of existence. Note that we are making no such invidious distinction. *All* parts of the meaning of a word are there only because the users of the word impute them; denotation and connotation (both kinds of connotation) are equally respectable, equally important.

This distinction between denotation and connotation is clear and important. But it is not simple. To complicate matters, a single word can have varying denotations in different contexts: for example, *scale* has different denotations in graphical representation, in zoology and in a grocer's shop; *wave* has different denotations in oceanography, in electronics and in a hairdresser's shop; *pavement* has different denotations in England and America.

Some words have a definition but no concrete denotation. We have no concrete referent for *infinity, polarity, brittleness* or *parameter.* The referent for each of these is a mental construct, an abstraction or a perception of a relationship; scope for varying interpretations of these terms is substantial. Many of the terms we use in mathematics are understood entirely by their defining or informative connotations; and some terms have different informative connotations or definitions in different contexts: for example, *plasma* in physics and biology, *base* in chemistry and architecture, and *stress* in engineering and psychology.

Above all, it is the affective or associative connotations that give rise to ambiguity — to what William Empson called[7] 'room for alternative reactions to the same piece of language'. Consider the words *adolescent, progressive, latent, humid, consistency, significance* and *efficiency.* Can we be confident that we all have identical associations for these words?

Usually, in scientific writing we want to minimize the possible variety of connotations, because we want to communicate about positive, concrete things — chemical substances, physical materials, units of machinery, instruments, reactions, states or conditions which must be carefully delimited. Scientists try to do this by using special vocabulary. W. E. Flood, in a study of the structure and meanings of scientific words[8], suggests that:

> the scientist needs suitable names by which to identify the various abstractions with which he deals — processes, states, qualities, relationships, and so on. Thus, after Faraday had investigated the passage of electric currents through different solutions and noticed the resulting liberation of chemical substances, the term *electrolysis* was invented. This one word was a kind of shorthand symbol for the process; it 'pinned down' the process and conveniently embraced its many aspects. From then on it was possible to think about the process and to talk about it to others.

But it is dangerous to follow Flood's simple theory and to accept that specially created words can subsequently be used with absolute confidence. He contends[8] that:

> a scientist avoids the ordinary words of the language; he prefers his own words. These words can then be rigorously defined and given the necessary precision of meaning.

It would be interesting to ask every reader to write a 'rigorous' definition of a selection of scientific and technical terms such as *calibration, hydraulic, lithography, matrix, parameter* and *transistor* to see with what agreement we 'pinned down' the processes, objects or states involved.

Our faith in the possibility of defining scientific terms rigorously must surely be shaken by Waismann's demonstration[9] that we can never remove all possible doubt about the meaning of words, even of terms denoting physical objects:

> Is there anything like an exhaustive definition that finally and once and for all sets our mind at rest? 'But are there not exact definitions at least in science?' Let's see. The notion

of gold seems to be defined with absolute precision, say by the spectrum of gold with its characteristic lines. Now what would you say if a substance was discovered that looked like gold, satisfied all the chemical tests for gold, whilst it emitted a new sort of radiation? 'But such things do not happen'. Quite so; but they *might* happen, and that is enough to show that we can never exclude altogether the possibility of some unforeseen situation arising in which we shall have to modify our definition. Try as we may, no concept is limited in such a way that there is no room for any doubt. We introduce a concept and limit it in *some* directions; for instance, we define gold in contrast to some other metals such as alloys. This suffices for our present needs, and we do not probe any farther. We tend to *overlook* the fact that there are always other directions in which the concept has not been defined. And if we did, we could easily imagine conditions which would necessitate new limitations. In short, it is not possible to define a concept like gold with absolute precision, that is in such a way that every nook and cranny is blocked against entry of doubt.

Waismann described words such as *gold* as being of an 'open texture'. He distinguishes carefully between 'open texture' and 'vagueness':

A word which is actually used in a fluctuating way (such as *heap* or *pink*) is said to be vague; a term like *gold*, though its actual use may not be vague, is non-exhaustive or of an open texture in that we can never fill up all the possible gaps through which a doubt may seep in.

In using many apparently reliable scientific terms, we are troubled by their 'vagueness'. Consider such terms as *liquid*, at an *elevated* temperature. We cannot tell to what temperature the term *elevated* applies. Presumably, minus 50°C and plus 500°C are outside the limits of this term; but just what *is* the central temperature that is *elevated*, and what ranges on either side of it represent more or less elevated temperatures? What is an *accurate* measurement, a *stable* flow-rate, or an *efficient* reaction?

Vagueness is not necessarily damaging. Many of our words are essentially generic, and usefully so. General words such as *vehicle*, *metal* or *reaction* are often properly used when a writer does not want to specify a *three-wheeled car, iron* or *esterification.* Our judgement of the suitability of the words must depend on whether or not the receivers are able to take from them as much information as the writer intended.

Vagueness is not only an inherent quality of many words. It is, as Willard Quine points out, also a natural consequence of the basic mechanism of word learning. We learn many of our words by observing the customary verbal response to certain stimulations. We learn by observation, imitation or instruction to use certain terms to describe certain objects or states, reactions or relationships. Our 'reward' for making the right response, or using the right word, is observably successful communication with others. Our difficulties arise when we are not entirely certain of the proper verbal signal to use in a given situation; our understanding of the term has 'fuzzy edges'[10].

The penumbral objects of a vague term are the objects whose similarity to ones for which the verbal response has been rewarded is relatively slight. Or, the learning process being an implicit induction on the subject's part regarding society's usage, the penumbral cases are the cases for which that induction is most inconclusive for want of evidence. The evidence is not there to be gathered, society's members having themselves had to accept similarly fuzzy edges when they were learning. Such is the inevitability of vagueness on the part of terms learned in the primitive way; and it tends to carry over to other terms defined on the basis of these.

Regrettably, though I suspect inevitably, scientists and engineers, especially those interpreting the ideas and the information of other writers or teachers, often find themselves learning words in the 'primitive' way; and many scientific terms are held in their understanding by means of other words which are themselves only dimly understood. There is little we can do other than try constantly to bring as many terms as possible out from the 'penumbra' into the brighter light of fuller understanding.

In other words, we have to realize that we can rarely achieve absolute precision of meaning such as would satisfy the philosophers. The best we can achieve is what (to use Waismann's expression) will suffice for our present needs, making it unnecessary to probe farther: and that must be interpreted as what will suffice in a given context to convey our information to our readers without requiring from them more than a normal commitment to disambiguation.

I hope this interpretation of 'what will suffice' would satisfy the text-book writers I quoted at the beginning, and Flood and Waismann. It is not question-begging or a shrugging-off of responsibility, for it does not suggest that we should go ignorant of the causes of our semantic difficulties. We must certainly be alert to the problems attached to establishing the meaning of *meaning*; but we should acknowledge that it is easier to demonstrate the incompleteness of existing theories of meaning than to devise a new theory that can be applied in all situations; and meanwhile we have to get on with communication as best we can! We can do so with reasonable success if we take care.

In my view, therefore, writers who seek advice on overcoming ambiguity must be made aware of semantic problems, but at the same time encouraged to adopt a pragmatic attitude. They must learn to identify what extent of knowledge, what range of vocabulary, what capacity to disambiguate, they can expect from various audiences. They will be in for some surprises. For example, they can expect three in every ten British first-year undergraduates in applied science departments to be unable to define *synthesis*[11]. At a more general level, they can expect four in every ten of the fifteen-year-olds leaving British secondary schools to be defeated by *circuit*[12]. But facts such as these will provide a useful balance to abstract semantic studies.

I should like to prescribe a compulsory piece of training for all embryo scientists and engineers. They should be required to write a description of a start-up procedure or a set of operating instructions or a routine to be followed in fault-finding. They should then be obliged to watch silently while an operator or maintainer tried to do the job solely in accord with the written instructions. This, I am sure, would teach as much about the right things to consider in avoiding ambiguity as several hours' reflection on abstruse aspects of semantics or philosophy.

References

1. Ehrlich E. and Murphy D., *The Art of Technical Writing,* Bantam Books, 1964, p.113.
2. Little P., *Communication in Business,* Longmans, 1965, p.41.

3. Gloag J., *How to Write Technical Books*, Allen and Unwin, 1950, p.30.
4. Graves H. F. and Hoffman L. S. S., *Report Writing*, Prentice-Hall, 1965, p.26.
5. Emberger M. R. and Hall M. R., *Scientific Writing*, Harcourt Brace, 1955, p.157.
6. Salomon L., *Semantics and Common Sense*, Holt Rinehart and Winston, 1966, p.14.
7. Empson W., *Seven Types of Ambiguity*, Penguin Books, 1961, p.1.
8. Flood W.E., *Scientific Words, Their Structure and Meaning*, Oldbourne, 1960, p.viii.
9. Waismann F., 'Verifiability', in *The Theory of Meaning*, ed. G. H. R. Parkinson, Oxford University Press, 1968, p.38.
10. Quine W. V. O., *Word and Object*, MIT Press, 1960, pp.125-126.
11. Kirkman A. J., 'Command of Vocabulary Among University Entrants', *Educational Research*, Newnes Ltd, Vol.9, No.4, 1967, p.151.
12. Flood W. E., *The Problem of Vocabulary in the Popularisation of Science*, Oliver and Boyd, 1957, p.87.

Appendix 1
Texts used in the survey made with the co-operation of the Institution of Chemical Engineers

Explanatory Notes

Six different versions of part of an engineering report are enclosed. You are invited to show on the accompanying card which of the versions you find most comfortable to read, easiest to grasp and simplest to digest. Please add any comments you would like to make on why you prefer some versions and reject others, and return the card to The Institution of Chemical Engineers.

The versions should be judged as part of a joint-authorship paper on *Liquid Flow in Packed Columns*. The paper describes the spread of liquid through a bed of randomly dumped packings in terms of rivulets, each of which has a stable but randomly oriented path through the bed.

Version T

In all the experiments in the present work it was found that the distributions were time-independent (this has been noted by most previous workers on liquid distribution) and this tends to indicate the existence of a stable flow pattern in the packed bed. It is thought to be of interest to determine to what extent this pattern of flow is dependent on bed structure and to what extent it is dependent on movement of a purely random nature of the initial liquid particles producing wetted paths through the packing. The effect on any given arrangement of packing of complete overload of the bed by high liquid flow-rate (pre-flooding) might be predicted to be the alteration of a flow pattern, dependent upon which of the paths through the packing experienced wetting, whereas the effect of re-packing or 'stirring' the bed will be changes in the structure of the bed. Liquid spread measurements utilizing salt tracer techniques were carried out both prior to and subsequent to either pre-flooding of the packed bed or effecting changes in the structure of the bed by stirring (½ in. rings were used), other conditions, packed height and liquid rate, being kept constant. It was apparent from the results that both pre-flooding and bed-stirring exerted a considerable influence on the stability of the flow pattern but that re-packing exerted a substantially higher degree of influence than pre-flooding, hence indicating that bed structure is the factor of greatest importance in the determination of flow pattern stability.

Version R

All the experiments showed the distributions to be time-independent. This has been noted by most previous workers on liquid distribution. It indicates the existence of a stable flow pattern in the packed bed. This flow pattern may be dependent on two factors. The first is bed structure, the second is purely random movement of the initial liquid particles producing wetted paths through the packing. It is of interest to determine to what extent flow pattern is dependent on these two factors. For a given arrangement of packing it could be expected that complete overloading of the bed by a high liquid flow-rate (pre-flooding) would alter a flow pattern. For the same arrangement, re-packing or 'stirring' the bed would be expected to cause changes in the structure of the bed. Liquid-spread measurements were undertaken to test this. They were carried out before and after either pre-flooding the packed bed or changing the bed structure by stirring. Half-inch rings were utilized for the latter purpose. Other conditions, namely packing height and liquid rate, were kept constant. Salt tracer experiments were used. Both pre-flooding and stirring influenced the stable flow pattern. The effect of re-packing, however, was substantially greater than that of pre-flooding. This means that stable flow pattern is determined mainly by bed structure.

Version Y

In all the experiments it was found that the distributions were time-independent (this has been noted by most previous workers on liquid distribution), and this indicates the existence of a stable flow pattern in the packed bed. It is of interest to determine to what extent this flow pattern is dependent on bed structure and to what extent it is dependent on a purely random movement of the initial liquid particles producing wetted paths through the packing. For a given arrangement of packing, completely overloading the bed with a high liquid flow-rate (pre-flooding) might be expected to alter a flow pattern, depending on which paths through the packing are wetted, whereas re-packing, or 'stirring' the bed, will cause changes in the structure of the bed. Liquid-spread measurements were made before and also after either pre-flooding the packed bed or changing the structure of the bed by stirring (½ in. rings were used), other conditions, packed height and liquid rate, being kept constant. Salt tracer experiments were used. From the results of the experiments it can be seen that both pre-flooding and stirring the bed influence the stable flow pattern, but that the effect of repacking is much greater than that of pre-flooding. That is, the stable flow pattern is determined mainly by the structure of the bed.

Version H

All our experiments showed that the distributions did not depend on time. This has been shown by most previous work on liquid distribution, and it suggests that the flow pattern in the packed bed is stable.

We thought it would be interesting to know to what extent this flow pattern depends on bed structure and to what extent it depends on the random movement of the initial liquid particles making wetted paths through the packing. In any given packing arrangement, re-packing or 'stirring' might be expected to change the structure of the bed; complete overloading of the bed with a high liquid flow-rate (pre-flooding) might be expected to alter a flow pattern, depending on which paths through the pattern were wetted.

So we used salt tracer methods to measure liquid spread before and after stirring (using ½ in. rings) and before and after pre-flooding. Other conditions — packed height and liquid rate — were kept constant. Both stirring and pre-flooding affected the stable flow pattern, but stirring had much greater effect than pre-flooding: obviously, the structure of the bed is mainly responsible for the stable flow pattern.

Version F

Time-independence of distributions, indicative of stable flow pattern in the packed bed, occurred in all experiments, confirming most previous liquid distribution work. The degree of dependence of this pattern on bed structure and/or on production of wetted paths through the packing by randomly moving particles of the initial liquid is of interest, as in any specified packing arrangement, complete bed overload by high liquid flow-rate (pre-flooding) might be expected to result in alteration of a flow pattern, dependent on which paths through the packing were wetted, while changes in bed structure might result from re-packing or 'stirring' the bed. Liquid-spread measurements by salt tracer techniques were made before and after either packed-bed pre-flooding or structure-changing by stirring (utilizing ½ in. rings); other conditions, packed height and liquid rate, being maintained constant. The effect of re-packing on flow pattern stability was far in excess of that of pre-flooding. Hence flow pattern stability is determined mainly by bed structure.

Version B

In all our experiments we found that our distributions were not dependent on time. This has been noted by the majority of our predecessors in work in the liquid distribution field, and we believe that this indicates the existence of a stable flow pattern in the packed bed.

We felt it would be of considerable interest if we were able to demonstrate in our experiments the extent to which this pattern depends on bed structure and to what extent it depends on a purely random movement of the initial liquid particles producing wetted paths through the packing. We might reasonably hypothesize that for any given packing arrangement, complete overloading of the bed by a high liquid flow-rate (known as pre-flooding) would alter our flow pattern, depending on which paths through the packing were wetted. We would expect, on the other hand, that re-packing or 'stirring' the bed would bring about changes in the fundamental structure of the bed.

We undertook measurements of liquid spread before and following either pre-flooding the packed bed or changing the structure of the bed by stirring (which we effected by means of ½ in. rings).   Other conditions, viz. packed height and liquid rate, we maintained at a constant level.   Salt tracer techniques were our choice for the experiments.

Our results reveal that both pre-flooding and stirring the bed exert a marked influence on the stable flow pattern; nonetheless, the effect of re-packing is of far greater magnitude.   We must conclude, therefore, that the stable flow pattern is determined mainly by the structure of the bed.

Appendix 2
Texts used in the survey made with the co-operation of the British Ecological Society

Explanatory Notes

Members of the British Ecological Society are invited to co-operate in an experiment aimed at establishing what style is most readable and acceptable for scientific reports and papers.

Six different versions of part of a scientific paper are enclosed. You are invited to show on the accompanying card which of the versions you find most comfortable to read, easiest to grasp and simplest to digest. Please add any comments you would like to make on why you prefer some versions and reject others, and return the card.

The versions should be judged as part of a joint-authorship paper for a professional journal such as *The Journal of Applied Ecology*. The paper describes work aimed at discovering which types of pasture are preferred for grazing by wigeon. Full nomenclature for the wigeon (*Anas penelope L.*) and for the species of grasses (*Festuca rubra L., Puccinellia maritima* (Huds) Parl, and *Agrostis stolonifera L.*) is given in the title of the paper.

There is no set order in which the versions should be read. Indeed, it is desirable that there should be as much variety in order of reading as possible. To achieve this, I suggest that each reader starts with the letter nearest to the initial letter of his surname.

For the sake of realism, the versions are based on genuine subject-matter reported in a published paper. I would stress, however, that none of these versions is the responsibility of the authors of the original paper!

Version S

Bird Island saltings comprise three abundant species of grass. The species are *Festuca rubra* (red fescue), *Puccinellia maritima* (salt-marsh grass) and *Agrostis stolonifera* (creeping bent grass). For the purpose of this study, two major types of sward were distinguished subjectively.   These were areas dominated by *Festuca rubra* and those by *Puccinellia maritima.* Of the area studied, approximately 59% was swards dominated by *Festuca rubra.* These swards also contained varying proportions of *Agrostis stolonifera* and a little *Puccinellia maritima.* Swards in the other 41% of the area were dominated by *Puccinellia maritima.* These swards also contained varying proportions of *Agrostis stolonifera* and a little *Festuca rubra.*

Observations were carried out from an observation tower overlooking the Island in the winter of 1970-71. It was observed that the degree of disturbance due to visitors to the area was of a sufficient order to inhibit wigeon from frequenting the Island during daylight hours.   However, droppings were observed on the Island, and this together with limited observations carried out at sunrise, sunset and on moonlit nights suggested the return of wigeon to graze at night.

It was decided that a count should be made of the numbers of wigeon feeding on the two main types of sward represented on the Island. This would enable a comparative assessment of the feeding preferences of the wigeon to be made.   Owen (1971) suggested that in view of the fact that European white-fronted geese (*Anser anser albifrons* (Scopoli)) defecate every 3½ minutes, a droppings count is a sensitive method of estimating grazing usage. The same basic assumption was used in the determination of feeding preferences in this work. It was assumed that the wigeon's quantitative defecation would be in proportion to the length of time spent grazing.

Counts were made in six plots in each of the two swards. Inter-count intervals were dictated by times when submergence of the Island by high spring tides made counting impossible. To avoid removal of the droppings by the tide, the first count was undertaken just prior to flooding. After each count, droppings were removed or trodden into the ground. Hence the possibility of duplication during ensuing counts was avoided. Each plot size was 10 × 3 m. Posts used as markers were small to minimize disturbance to the wigeon by alien objects on the feeding ground.

Selection of the six plots in each sward was based on an initial visual differentiation between the two swards. Due to morphological differences, the naked eye could readily detect

differences between the swards. The swards in which the plots
were placed were later compared by a 1/20 m² (25 × 20 cm)
quadrat sampling technique. Percentage cover estimates were
made for the three grass species in twenty randomly selected
quadrats in each plot. To test for changes to the sward consequent upon wigeon grazing, the percentage frequency occurrence of various grass blade lengths of the species in a *Puccinellia/
Agrostis* plot was determined. Two determinations were carried
out at a one month interval, at times when maximum wigeon
numbers were present. A full description of the point intercept method utilized in the determinations is given in Appendix
1.

Version B

The saltings on Bird Island have three abundant species of
grass, namely *Festuca rubra* (red fescue), *Puccinellia maritima*
(salt-marsh grass) and *Agrostis stolonifera* (creeping bent grass).
Approximately 59% of the study area was covered by swards
in which *Festuca rubra* predominated with varying proportions
of *Agrostis stolonifera* and a little *Puccinellia maritima*. The
remaining area consisted mainly of *Puccinellia maritima* but
with a little *Festuca rubra,* and again with varying proportions
of *Agrostis stolonifera*. For the purpose of this study two
major sward types were distinguished subjectively, namely areas
dominated by *Festuca rubra* and those by *Puccinellia maritima*.
    In order to carry out a comparative assessment of the feeding
preferences of the wigeon, it was decided that a count should
be made of the numbers feeding on the two main types of
sward represented on the Island. Owen (1971) suggested that
as the European white-fronted geese (*Anser anser albifrons*
(Scopoli)) defecate every 3½ minutes, counting the droppings
is a sensitive method for estimating grazing usage. The same
basic presumption, namely that the wigeon would defecate
quantitatively in proportion to the length of time spent grazing,
was used in the determination of feeding preferences in this
work. Equal numbers of plots were marked out in both swards.
Observations carried out from an observation tower overlooking
the Island in the winter of 1970-71, however, showed that the
degree of disturbance caused by visitors to the area was sufficient to keep the wigeon off the Island during the day. The

presence of droppings on the Island and limited observations carried out at sunrise, sunset and moonlit nights indicated that the wigeon were returning to graze at night.

As this study was intended to compare the preference of wigeon for the two types of sward on Bird Island, the number of droppings was counted on three and four occasions during the period November-March 1971/72 and 1972/73 respectively in six plots in each of the two swards. The interval between counting was governed by the times when the Island was covered at high spring tides. The first count was just before flooding, as otherwise the droppings would have been washed away. The droppings were removed or trodden into the ground following each count in order to prevent duplication during ensuing counts. Each plot was 10 × 3 m and was marked with small posts in order to reduce disturbance of the wigeon by the presence of alien objects on their feeding ground.

The selection of the sites of the six plots in each sward was made on an initial visual differentiation between the two swards which owing to their different morphology were quite distinct to the naked eye. The swards in which the plots were placed were later compared by a $1/20$ m^2 (25×20 cm) quadrat sampling technique, in which percentage cover estimates were made for the three grass species in twenty random quadrats in each plot. In order to test for changes to the sward as a result of wigeon grazing, the percentage frequency occurrence of the various blade lengths of the grass species in a *Puccinellia/Agrostis* plot was determined by a point intercept method on two occasions at a month's interval, when maximum numbers of wigeon were present. This intercept method is described in full in Appendix 1.

Version M

Three sorts of grass grow abundantly on Bird Island. They are red fescue (*Festuca rubra*), salt-marsh grass (*Puccinellia maritima*) and creeping bent grass (*Agrostis stolonifera*). About 59% of our study area was covered by swards which were mainly red fescue, with varying amounts of creeping bent grass and a little salt-marsh grass. The rest was mainly salt-marsh grass, with varying amounts of creeping bent grass and a little red fescue. For the purpose of our study, we subjectively distinguished between two major sward types: one dominated by the fescue and one dominated by the marsh grass.

As it was our intention to assess which type of grass the wigeon preferred for feeding, we decided to count the numbers grazing on our two main types of sward. We knew that Owen, in 1971, had suggested that as European white-fronted geese (*Anser anser albifrons* (Scopoli)) defecate every 3½ minutes, a count of the droppings gives you a good idea of the extent to which the birds are using the grazing. So we worked on the same assumption: that is, we took it that we could base our assessment of feeding preferences on the idea that the number of wigeon droppings would be proportional to the time they spent grazing.

We marked out the same number of plots in each sward. We knew from observations we had made in the winter of 1970-71 that visitors to the area scared the birds away from the Island during the day. But we also knew that the birds were coming back to graze at night, because we found droppings on the Island, and we kept watch a few times at sunrise, sunset, and on moonlit nights.

Our idea being to see which type of sward the wigeon preferred, we counted the number of droppings in six plots in each of the two swards on seven occasions in all — three in November-March 1971-72 and four in November-March 1972-73. Of course, we could not count when the Island was under water during the high spring tides. The first count was just before the island was flooded, when the droppings would have been washed away. After each count, we always took the droppings right away or else trampled them well into the ground, so that we would not count the same droppings again during the next count. The plots were 10 × 3 m in size, and were marked with posts — *small* posts, so that the wigeon would be disturbed as little as possible by strange things in their feeding grounds.

At first, we chose the sites for the six plots in each of the two swards simply by looking at them: they were plainly different in form. Later, we compared them by a 25 × 20 cm quadrat sampling technique, to get an estimate of the percentage cover for our three grass species. This was done in twenty random quadrats in each plot. To test for changes in the sward after the wigeon had grazed, we used a point intercept method to check the percentage frequency of the lengths of blades of grass in a salt-marsh-dominated plot. We did this twice, with a month in between, at times when the largest number of wigeon were present.   Full details of our intercept method are in Appendix 1.

Version Y

Three species of grasses grow abundantly on Bird Island saltings: red fescue, salt-marsh grass and creeping bent grass (*Festuca rubra, Puccinellia maritima* and *Agrostis stolonifera*).   For our study, we distinguished subjectively between two main types of sward: swards dominated by red fescue and swards dominated by salt-marsh grass.   Approximately 59% of the study area was dominated by red fescue, with a little salt-marsh grass and varying proportions of bent grass.   The remaining 41% was dominated by salt-marsh grass, with a little red fescue and varying proportions of bent grass.

In the winter of 1970-71, observations from a tower overlooking the island showed that visitors to the area caused enough disturbance to keep the wigeon off the island during the day.   However, the presence of droppings on the island, and a few observations at sunrise, sunset and on moonlit nights indicated that the wigeon were returning to graze at night.

Our aim was to assess feeding preferences by counting the numbers of wigeon feeding on the two types of sward.   In 1971, Owen suggested[1] that as European white-fronted geese (*Anser anser albifrons* (Scopoli)) defecate every 3½ minutes, a count of their droppings gives a sensitive estimate of the extent of their grazing.   We decided to make the same assumption: that the number of droppings would show how long the wigeon spent grazing, and thus which grass they preferred. Accordingly, we marked out equal numbers of plots in both types of swards. Each plot was 10 × 3 m.   The posts marking the plots were small, to minimize disturbance of the birds by strange objects on their feeding grounds.

The droppings in six plots in each sward were counted on three occasions in November-March 1971-72 and on four occasions in November-March 1972-73.   The intervals between counts were governed by the high spring tides, which covered the island.   The first count was just before flooding, which would have washed away the droppings. After each count, the droppings were removed or trodden into the ground, to prevent duplication in subsequent counts.

As the swards were visibly different in form, the sites of the six plots in each sward were selected subjectively at first. Later, the swards containing the plots were compared by a 25 × 20 cm quadrat sampling technique, in which percentage cover was estimated for the three species of grasses. Twenty quadrats were sampled at random in each plot.

To test whether grazing by wigeon changed the sward, we calculated the percentage frequency of blade lengths of the grasses in a plot dominated by salt-marsh grass. The calculation was made on two occasions, one month apart, when the maximum number of wigeon were present. The point intercept method used is described fully in Appendix 1.

Version F

Three grass species, *Festuca rubra* (red fescue), *Puccinellia maritima* (salt-marsh grass), and *Agrostis stolonifera* (creeping bent grass) occur abundantly on Bird Island saltings. For this study, two major sward types (*Festuca*-dominated, containing varying amounts of *Agrostis stolonifera* and a small quantity of *Puccinellia maritima*, and *Puccinellia*-dominated, containing varying amounts of *Agrostis stolonifera* and a small quantity of *Festuca rubra*) were distinguished subjectively, comprising 59% and 41% of the study area respectively.

Observations from an observation tower overlooking the Island in winter 1970-71 indicated an incidence of disturbance due to visitors sufficient to cause diurnal absence of wigeon from the Island. However, limited observations at sunrise, sunset, and on moonlit nights, and presence of droppings on the island suggested nocturnal return and resumption of grazing.

Comparative assessment of wigeon feeding preferences by determination of numbers feeding on the Island's main sward types was carried out by counting of droppings. Following Owen's (1971) theory that since European white-fronted geese (*Anser anser albifrons* (Scopoli)) defecate at 3½ minute intervals, a sensitive measure of grazing usage may be obtained from a droppings count, it was decided to accept number of wigeon droppings as a function of time spent grazing. In both swards, demarcation of equal numbers of 10 × 3 m plots was accomplished by utilization of small posts to reduce wigeon disturbance effects due to presence of alien objects on the feeding ground.

To compare feeding preferences, a count was made of numbers of droppings in six plots in each of the two sward types on three and four occasions during November-March 1971-72 and 1972-73 respectively, the interval between counts being governed by immersion of the Island at the time of high spring tides. Thus the initial count took place immediately prior to flooding to avoid removal of droppings by water. Duplication during ensuing counts was prevented by removal of droppings or treading into the ground.

Site selection of the six plots in each sward was based on an initial visual differentiation between the sward types, whose heterogeneity in respect of morphology was macroscopically apparent. A comparison of swards in which plots were located was however later performed utilizing a 25 X 20 cm quadrat sampling technique, enabling percentage cover estimation for the three grass species in twenty randomly selected quadrats in each plot. As a test for sward modifications consequent upon wigeon grazing, a point intercept method was employed (for full details see Appendix 1) on two occasions at an interval of one month, at peak wigeon population levels, to determine percentage frequency occurrence of various blade lengths of grass species in a *Puccinellia/Agrostis* plot.

Version R

In the saltings in question three abundant grass species occur, namely *Festuca rubra* (red fescue), *Puccinellia maritima* (salt-marsh grass), and *Agrostis stolonifera* (creeping bent grass). Of the area concerned in the study, of the order of 59% was covered by swards in which grass of the *Festuca rubra* type was predominant, with proportions of *Agrostis stolonifera* at varying levels, and also a small amount of *Puccinellia maritima*, whilst in the remaining 41% (approximately) *Puccinellia maritima* was predominant, with a small amount of *Festuca rubra*, and again proportions of *Agrostis stolonifera* at varying levels. In the current study, two principal types of sward were distinguished initially for experimental purposes on a subjective basis, namely areas dominated by *Festuca rubra* and those dominated by *Puccinellia maritima*.

It was thought that a comparative assessment of the feeding preferences of the wigeon could be achieved by observation of the numbers engaged on feeding on the two main types of sward. A previous worker (Owen, 1971) had suggested that as the European white-fronted geese (*Anser anser albifrons* (Scopoli)) defecate at 3½ minute intervals, grazing usage is capable of accurate measurement by the method of counting droppings. A similar premiss was postulated in this work: that a quantitative measure of defecation would be directly proportional to the total length of time spent grazing by the wigeon, and from this feeding preferences could be determined. Within each of the principal types of sward distinguished on the Island, equal numbers of plots were identified with markers. From observations carried out from an observation tower located in a position overlooking the Island in the winter of 1970-71, it was apparent that the degree of disturbance caused by the presence of visitors to the area was sufficient to cause the wigeon to refrain from frequenting the Island in the course of the day. On the other hand, from the presence of droppings on the Island and from a limited number of observations undertaken at sunrise, sunset and moonlit nights it was apparent that the birds were returning at night to resume their grazing activity.

As it was the intention of the current study to perform a comparison of the preferences demonstrated by the wigeon between the two principal types of sward on Bird Island, a count was made of the number of droppings on three and four occasions during the period between November-March 1971-72 and November-March 1972-73 respectively in each of six plots in each of the two sward types. The factor governing the interval between counting was the times during which immersion of the Island by high level spring tides was complete. The initial count was carried out immediately prior to flooding as otherwise removal of the droppings by washing action of the water would have taken place. Following each count, droppings were removed or trodden firmly into the ground, thus preventing duplication during counts on subsequent occasions. Each plot was 10 X 3 m in area and was marked with small posts in order to diminish the disturbance to the wigeon occasioned by the presence of alien objects in their customary feeding ground.

Site selection for the six plots in each sward type was based on an initial differentiation between the two swards made visually, the swards being quite distinct to the unaided eye due to their morphological distinctiveness. A comparison of the swards in which the plots were placed was subsequently performed by means of a $1/20 \text{ m}^2$ (25×20 cm) quadrat sampling technique, in which percentage cover estimates were made in twenty quadrats chosen at random in each plot for each of the three grass species under review. To identify changes in the sward brought about by the grazing of the wigeon a test was carried out to determine by a point intercept method the percentage frequency occurrence of blades of varying lengths of the grass species occurring in a *Puccinellia/Agrostis* plot on two occasions separated by a one month interval, when maximum numbers of wigeon were present. Full details of the intercept method concerned are shown in Appendix 1.

Appendix 3
Texts used in the survey made with the co-operation of the Biochemical Society

What is Good Style for Scientific Writing?

Members of the Biochemical Society are invited to co-operate in an experiment aimed at establishing what style is most readable and acceptable for scientific reports and papers.

Six different versions of part of a scientific paper follow. You are invited to show on the attached slip which of the versions you find most comfortable to read, easiest to grasp and simplest to digest. Please add any comments you would like to make on why you prefer some versions and reject others, and return the slip.

The versions should be judged as part of a joint-authorship paper for a professional journal such as the *Biochemical Journal*. The paper describes an investigation of how the secretion of growth hormone from anterior pituitary of rats is affected by cyclic AMP. The extract is taken from the introductory section of the paper.

There is no set order in which the versions should be read. Indeed, it is desirable that there should be as much variety in order of reading as possible. To achieve this, each reader is asked to start with the letter nearest to the initial letter of his surname.

For the sake of realism, the versions are based on genuine subject-matter reported in a published paper. However, none of these versions is the responsibility of the authors of the original paper!

Version A

This study was to devise a simple system for further investigation *in vitro* of control of growth hormone secretion. It was to be based on measurement of growth hormone release rate from isolated fragments of rat anterior pituitary. Rat pituitary was chosen because of its ready availability in the fresh state. It is also relatively small. This allows fragments to be obtained with minimal tissue damage. The investigation used the double antibody radioimmunoassay technique recently developed in the Biochemistry Department of the University of Wessex (Black & White, 1970). This procedure is rapid, sensitive and specific for rat growth hormone assay. It has been shown to be valid when applied to the determination of the hormone released into media in which rat anterior pituitary fragments have previously been incubated (Lavender, Greenhedge, Hawthorne & Berry, 1972).

A preliminary account of some of the findings reported here has previously been published (Brown & Green 1973). That account included a demonstration of a direct effect of dibutyryl cyclic AMP in stimulating growth hormone secretion *in vitro*. The present paper shows that rat pituitary fragment preparation is a reproducible and sensitive system for *in vitro* investigation of regulation of growth hormone release. Also, it is demonstrated that it is possible to distinguish between true secretion and hormone leakage. Hormone leakage, it is presumed, occurs due to cell damage during isolation of gland fragments. The system has been employed to study growth hormone secretion. It has also been used in the investigation of the metabolic requirements of the gland when active secretion is occurring in response to stimulation by cyclic AMP.

Version G

We set out to produce in this work a simple *in vitro* system which we could use to look further into the ways in which secretion of growth hormone can be controlled. We based our method on measurement of the rate at which growth hormone is released from isolated fragments of anterior pituitary from rats. We chose to use rat pituitary because it is easy to get fresh supplies and it is relatively small. This meant we could obtain fragments with only a little tissue damage.

In our work, we used the rapid, sensitive and specific double-antibody technique for radioimmunoassay of growth hormone in rats, developed recently in the University of Wessex (1). We knew it had been shown to be valid when used to check on the amounts of hormone released into media in which fragments of rat anterior pituitary had been incubated previously (2).

We have published some of our findings previously (3), including our demonstration of the direct effect dibutyryl cyclic AMP has on the stimulation of growth hormone secretion *in vitro*. In this paper, we present our evidence to support the concept that rat pituitary fragment preparation gives a reproducible and sensitive system for *in vitro* investigation of the regulation of growth hormone release. Also, we describe how, by using it, we have been able to distinguish between hormone leakage, which we presume occurs because of damage to cells during isolation of gland fragments, and true secretion. We have used the system to study growth hormone secretion, and we have also used it to investigate what a pituitary gland's metabolic requirements are when it is in an actively secreting state, responding to stimuli from cyclic AMP.

Version O

This study aimed at devising a simple *in vitro* system for further growth hormone secretion control mechanism investigations, based on growth hormone release rate measurements from isolated rat anterior pituitary fragments. Choice of rat pituitary was due to its relatively small size, thus permitting fragment section with minimal tissue damage, and ready availability in the fresh state. The University of Wessex's recently developed rapid, sensitive, specific, double-antibody radioimmunoassay technique (Black & White, 1970), whose validity has been demonstrated in application to determination of hormone release into media previously utilized for incubation of rat anterior pituitary fragments (Lavender, Greenhedge, Hawthorne & Berry, 1972) was used in these investigations. Preliminary findings, including demonstration of a direct effect of dibutyryl cyclic AMP on growth hormone secretion stimulation *in vitro*, have been published (Brown & Green, 1973). This paper shows rat pituitary fragment preparation to provide a reproducible and sensitive system for *in vitro* investigation of growth hormone release regulation, usable to distinguish between hormone leakage (occurring presumably from cell damage during gland fragment isolation) and true secretion. Study of growth hormone secretion and investigation of the metabolic conditions necessary for cyclic AMP-stimulated secretory processes to occur actively in the gland have been performed with this system.

Version W

It was the purpose of the present study to devise a simple system *in vitro*, based on measurement of the rate of growth hormone release from isolated fragments of rat anterior pituitary, which would be suitable for the further investigation of the mechanisms by which growth hormone secretion may be controlled. The choice of rat pituitary was made for these studies on grounds of its ready availability in the fresh state and its relatively small size which allows fragments to be obtained with minimal tissue damage. The investigations described below were carried out using the rapid, sensitive and specific double-antibody radioimmunoassay technique for rat growth hormone measurement recently developed in the Biochemistry Department of the University of Wessex (Black & White, 1970) which has been shown to be valid when applied to the determination of the hormone released into media in which rat anterior pituitary fragments have previously been incubated (Lavender, Greenhedge, Hawthorne & Berry, 1972).

A preliminary account of some of the findings reported here, including the demonstration of a direct effect of dibutyryl cyclic AMP in stimulating growth hormone secretion *in vitro*, has previously been published (Brown & Green, 1973). In the present paper, evidence is presented to show that the rat pituitary fragment preparation provides a reproducible and sensitive system for the investigation *in vitro* of the regulation of growth hormone release and that, by its use, it is possible to distinguish between hormone leakage, occurring presumably from cells damaged during isolation of the gland fragments, and true secretion. The system has been employed to study growth hormone secretion and to investigate the metabolic requirements exhibited by the gland when it is stimulated into an actively secreting state by cyclic AMP.

Version P

The purpose of the study reported here was the establishment of an unsophisticated *in vitro* system, based on measurement of the rate of release of growth hormone from isolated fragments of anterior pituitary from rats, which would provide a suitable technique for further elucidation of the mechanisms by which regulation of the growth hormone secretory process in the cells of the anterior pituitary may be achieved. The choice of rat anterior pituitary for these studies was made by virtue of the ready availability of this gland in the fresh state and its relatively small size which permits fragments being obtained with minimal damage to the tissues involved.

The investigations described in the following pages were carried out utilizing the rapid, sensitive, specific, double-antibody radioimmunoassay technique for rat growth hormone measurement recently developed in the Biochemistry Department of the University of Wessex (Black & White, 1970). That such a process has validity has been demonstrated by its application to the determination of hormone release into media in which incubation of rat anterior pituitary fragments has previously been carried out (Lavender, Greenhedge, Hawthorne & Berry, 1972).

A preliminary publication (Brown & Green, 1973) presented an account of some of the experimental findings reported below in this paper, including evidence demonstrating the direct role of dibutyryl cyclic AMP in the stimulation of secretion of growth hormone *in vitro*.

The present paper presents evidence to demonstrate that a sensitive and reproducible system for the *in vitro* investigation of the growth hormone release regulatory mechanisms is provided by the rat pituitary fragment preparation. Utilizing this system, distinction is possible between hormone leakage — occurring, it is presumed, due to cell damage during isolation of gland fragments — and true secretion. Studies of growth hormone secretion have been carried out employing this system, and also investigations of the metabolic requirements exhibited by the gland when stimulation to an actively secreting state by cyclic AMP is occurring.

Version J

Our aim was to devise a simple system for further *in vitro* investigation of ways of controlling secretion of growth hormone. The investigation was to be based on measuring the rate at which growth hormone is released from isolated fragments of anterior pituitary from rats. We chose rat pituitary because it is relatively small, so fragments can be obtained with only slight tissue damage. Also, fresh pituitary is easy to get. In the investigation, we used the rapid, sensitive and specific technique developed at the University of Wessex[1] for double-antibody radioimmunoassay of growth hormone from rats. This assay has been shown[2] to be valid when used to calculate how much hormone is released into media in which fragments of anterior pituitary from rats have been incubated.

Some of our findings, including the direct stimulation of growth hormone secretion by dibutyryl cyclic AMP *in vitro,* have been published previously[3]. In this paper, we show that a preparation of fragments from rat pituitaries is a reproducible and sensitive system for investigating *in vitro* how the release of growth hormone can be regulated. Using this preparation, it is possible to distinguish between true secretion and hormone leakage (which presumably occurs because cells are damaged when fragments of gland are isolated). We have used the preparation to study the secretion of growth hormone, and to investigate the metabolic conditions required if the pituitary is to secrete actively in response to cyclic AMP.

Appendix 4
Results and comments

How acceptable is the more readable style I advocate? According to my surveys, very acceptable. In all three surveys, the respondents voted clearly in favour of direct, active writing, with a minimum of specialist vocabulary, a judicious mixture of personal and impersonal constructions, short and uncomplicated sentences and liberal paragraphing. In all three surveys, a reasonably well written version in 'traditional' passive, impersonal style came a poor second. The majorities were 28%, 41% and 18·5%. The full results were as shown in Tables 1-3, pages 125-7.

I readily acknowledge a point made by many of the scientists who responded to my surveys. They pointed out that my samples of readers' views would be biased — that I should get replies from people who care about skilful writing but not from those who do not. I acknowledge that weakness; it is inherent in any surveys that rely on voluntary responses, especially voluntary responses by post. But I believe it is better to have these samples than to have no real evidence about scientists' opinions on writing. And I would suggest that the opinions of those who care about skilful writing are the ones we should value.

It is important to acknowledge, too, that we are unlikely ever to get unanimity about the readability of writing, as judgement is entirely subjective. I must repeat that the readability of a text is affected by many factors besides the choice of language. Obviously, the reader's background knowledge of the subject affects his capacity to comprehend what is being said. The intrinsic difficulty of the concepts being expressed is important. The reader's motivation is also significant: does he *want* to read the text or is he being forced to? Does he want to scrutinize the text closely or is his aim merely to skim it superficially? To make a full analysis of a reader's response to a text, we should take into account a wide range of psychological and even physical elements in the context in which the reading is taking place.

Table 1
Results of the survey made with the co-operation of the Institution of Chemical Engineers

Version		No. of times chosen as 'best'	
H	Direct, verbs mainly active, minimum of special vocabulary, judicious use of personal and impersonal constructions, sentences of varied length but mainly short and not complex, 3 paragraphs	706	(46%)
Y	Reasonably direct, verbs mainly passive, some roundabout phrasing to avoid personal constructions, reasonable amount of special vocabulary, sentences of varied length but some long, involved constructions, 2 paragraphs	280	(18%)
R	Reasonably direct, verbs mainly passive, some roundabout phrasing to avoid personal constructions, moderate use of special vocabulary, sentences almost all short, consequent bitty effect, 1 paragraph	245	(16%)
F	Reasonably direct, verbs mainly passive, much roundabout phrasing to avoid personal constructions, consequent extensive use of abstractions, much use of nouns as adjectives, sentences often long and too tightly packed with information, 1 paragraph	133	(9%)
B	Chatty but clear, verbs mainly active, little special vocabulary, some colloquial expressions, excessive use of personal constructions, sentences of varied length but mainly short and of simple structure, 4 paragraphs	121	(8%)
T	Roundabout and woolly, verbs mainly passive, excessive use of special vocabulary, much roundabout phrasing, sentences often long and involved, 1 paragraph	50	(3%)
		1535	(100%)

Table 2
Results of the survey made with the co-operation of the British Ecological Society

Version		No. of times chosen as 'best'	
Y	Direct, verbs mainly active, minimum of special vocabulary, judicious use of personal and impersonal constructions, sentences of varied length but mainly short and not complex, 6 paragraphs	302	(57·4%)
B	Reasonably direct, verbs mainly passive, some roundabout phrasing to avoid personal constructions, reasonable amount of special vocabulary, sentences of varied length but mainly short and not complex, 4 paragraphs	86	(16·4%)
F	Reasonably direct, verbs mainly passive, much roundabout phrasing to avoid personal constructions, consequent extensive use of abstractions, excessive use of special vocabulary, sentences often long and too tightly packed with information, 5 paragraphs	56	(10·7%)
S	Reasonably direct, verbs mainly passive, some roundabout phrasing to avoid personal constructions, too much special vocabulary, much use of nouns as adjectives, sentences almost all short, consequent bitty effect, 5 paragraphs.	46	(8·7%)
M	Conversational but clear, verbs mainly active, little special vocabulary, some colloquial expressions, excessive use of personal constructions, sentences of varied length but mainly short and of simple structure, 5 paragraphs	30	(5·7%)
R	Roundabout and woolly, verbs mainly passive, excessive use of special vocabulary, much roundabout phrasing, sentences often long and involved, 4 long paragraphs	6	(1·1%)
		526	(100%)

Table 3
Results of the survey made with the co-operation of the
Biochemical Society

Version		No. of times chosen as 'best'	
J	Direct, verbs mainly active, minimum of special vocabulary, judicious use of personal and impersonal constructions, sentences of varied length but mainly short and not complex, 2 paragraphs	283	(38·2%)
W	Reasonably direct, verbs mainly passive, some roundabout phrasing to avoid personal constructions, reasonable amount of special vocabulary, sentences of varied length but some long, involved constructions, 2 paragraphs	146	(19·7%)
G	Conversational but clear, verbs mainly active, little special vocabulary, some colloquial expressions, excessive use of personal constructions, sentences of varied length but mainly short and of simple structure, 3 paragraphs	137	(18·5%)
A	Reasonably direct, verbs mainly passive, some roundabout phrasing to avoid personal constructions, moderate use of special vocabulary, sentences almost all short, consequent bitty effect, 2 paragraphs	104	(14·0%)
P	Roundabout and woolly, verbs mainly passive, moderate use of specialized vocabulary, much verbose phrasing, sentences often long and involved, 4 paragraphs	37	(5·0%)
O	Reasonably direct, verbs mainly passive, much roundabout phrasing to avoid personal constructions, consequent extensive use of abstractions, much use of nouns as adjectives, sentences often long and too tightly packed with information, 1 paragraph	34	(4·6%)
		741	(100%)

This survey aimed only to show the influence of a number of linguistic features on the readability of a text. I wanted to stress the difference that can be made by choosing language that is active rather than passive, concrete rather than abstract, and 'everyday' rather than 'special'; to show that the digestibility of statements is affected by both the length and the complexity of the sentences used to express them; to show the importance of using a variety of sentence types, some personal and some impersonal, instead of artificially restricting the choice to one or the other; and to stress the importance of breaking the whole statement (that is, the whole text) into units which give the reader the information he needs in convenient order and in digestible 'bites'.

Clear majorities of readers agreed which tactical choices produced the best effect. Indeed, many readers expressed views about the Versions that were less polite than my thumbnail sketches in the tables of results. For example, comments on Version F of the 'wigeon' texts included:

— heavy, laborious and ponderous
— complex sentence structure
— sense obscured by the complicated sentence construction
— sentences too long and complicated
— short, but style is jargon, not very explicit and difficult to read
— style very bad, meaning unclear
— contains an excessive proportion of Latin derived polysyllables and apposition nouns in substitution for adjectives for either ready comprehensibility or acceptability
— dreadful jargon
— the deliberate use of jargon and nouns in apposition is excruciating English at its Germanic worst
— too much use of the passive
— clarity sacrificed for brevity
— almost unreadable
— only after reference to the others was comprehension complete
— practically incomprehensible
— is a caricature
— technical narcissism, i.e. writing for the writer's sake, not the reader's
— the commonest style that scientists use
— would probably be accepted for a learned journal, I am

afraid to say
- unfortunately this kind of writing gets into print on occasion
- I've chosen F as worst mainly because the style is so common in journals
- the booby prize...[but]...even F is reasonably lucid compared to some recent articles in the *Journal of Ecology*

In contrast, 10·7% of the respondents in the same survey felt that Version F was the *best*:

- concise, precise and to the point
- best: concise, logical progression of ideas, good organization makes it digestible
- concise, fully informative, least ambiguous or vague statement
- no surplus words
- preferred because it is the most succinct
- crisp, clear, easily understood
- stands well above the others

Plainly, judgements are made on more than linguistic features alone. The influence of custom and expectation is strong. This was demonstrated in respondents' reactions to Version M of the 'wigeon' texts. Most readers (including me) found that version irritating and distastefully self-congratulatory. However, it is not the *use* but the *abuse* of personal constructions that is to blame. Too many 'we's', combined with some colloquial expressions and a general looseness of construction, led to the view that version M was too 'chatty' and lacking the formality due in a scientific paper.

Many respondents were surprised and intrigued by their own reactions to Version M. Their views are summed up in these quotations:

- I am not sure why I object to the very personal and conversational approach in M. It is otherwise good
- clear, but I have inbuilt bias against 1st person writing in journals
- while version M is attractive, easily read, would it ever pass an editor?
- this version went in best of the lot, but was far too matey to avoid irritating me
- good television script
- the rather chatty style of M seems to get the idea across much quicker

- too self-consciously simple for the type of journal; fine for the layman although the meaning is clear
- childish though easily read
- clear but a trifle juvenile
- is quite childish but is very easily understood
- God preserve us from M!
- although readable, is irritatingly personal and chatty
- my objections to M are probably the result of a conservative reaction to the journalistic tone. The style seems more appropriate for spoken presentation but the information is presented clearly and directly
- I suspect I did not rate (M) highly because it is an unusual way of writing scientific results
- a bit folksy and naive, but clear and readable
- is really unacceptable in a scientific journal but does communicate fairly economically
- annoying for a *scientific* publication, although highly suitable for a talk or a semi-popular article
- comes over more as an *oral* presentation
- would be first or second preference if I were *listening* to the paper
- this version says it all very clearly but the lapses into journalese/American come as a shock and distract from the feeling of scientific objectivity
- very clear, but chatty style feels wrong; excellent for verbal delivery
- possibly alright for a talk to a natural history soc. but not for a published paper

These repeated assessments of M as very clear but not acceptable in a formal scientific paper accentuate the weight that must be given to linguistic propriety. They express objections that have nothing to do with accuracy or clarity. They carry at least a tinge of scientific snootiness. The facts in Version M are the same as the facts in all the other versions, and there is no suggestion that those facts in themselves are unworthy of scientific attention. So it must have been the language chosen in Version M that made many readers reject it as lacking due scientific dignity.

To identify this phenomenon is not to belittle it. It is a phenomenon to be found in most communication contexts, not just in science. I have drawn attention to it here because it seems to me important that we should understand the bases of our judgements about the acceptability of attempts at communication. In discussion of Version M we are brought face-to-face with the differences between what is acceptable expression in speech and what is acceptable expression in writing. If we are to become skilled presenters of scientific information, we must become sensitive to such differences.

Are scientists and engineers, on average, skilled presenters of scientific and technical information? Is there need for advice on style such as is offered in this book? Certainly, many respondents to my surveys thought so. Let me answer with one quotation from each survey:

it is certainly time that biologists gave more attention (and had their attention drawn) to the material they publish. *Most* of it is so tedious, impenetrable or boring that I am sure one of the main reasons why fewer students are willing to study some of the biological sciences is the wretched literature they have to read!

(British Ecological Society reader)

from extensive experience as an editor, I can say that if the average standard of writing of papers submitted for publication were to rise to that of *any* of the examples given, I would think the millenium had arrived.

(Biochemical Society reader)

Please make as much of this exercise as you can — it would save me countless hours if pseudo-intellectual gibberish were cast out.

(Institution of Chemical Engineers reader)